Disclaimer

The publisher of this book is by no way associated with the National Institute of Standards and Technology (NIST). The NIST did not publish this book. It was published by 50 page publications under the public domain license.

50 Page Publications.

Book Title: Emergency First Responder Respirator Thermal Characteristics: Workshop Proceedings

Book Author: Amy E. Mensch; Nelson P. Bryner

Book Abstract: The purpose of this workshop was to identify performance needs and establish research priorities to address the thermal characteristics of respiratory equipment used by emergency first responders. The workshop provided a forum for representatives from the first responder community, self contained breathing apparatus (SCBA) and component manufacturers, and research and testing experts to discuss issues, technologies, and research associated with SCBA high temperature performance. The goals of the workshop were defined in two parts: 1) Clarify baseline information, including the current state-of-the-art, applicable fire service events, and current related research, and 2) Research planning, including identification of performance needs and short and long term research priorities. Presentations were given to explain the current SCBA and certification process, understand experience from actual fire service incidents, and review the current state of respirator research. After the presentations, the workshop divided into three working group sessions to discuss performance needs and research priorities in smaller groups. Suggested topics for discussion included: a) Current Equipment, b) Current Practice and Usage, c) Future Trends, d) Short Term Research Needs, e) Long Term Research Needs, and f) other issues. The results of the three smaller groups' deliberations were discussed when the full workshop reconvened. The responses from each group were merged into a combination of issues that related to the use and performance of the lens of the SCBA. The primary concerns and research priorities were the characterization of the fire fighter environment, performance of current and new technology, development of representative and realistic testing, and improvements to fire fighter training on the limitations of protective equipment. A significant amount of discussion concentrated on the testing for NFPA certification, which currently contains limited thermal testing.

Citation: NIST SP - 1123

Keyword: fire fighting; first responder; heat detection; heat flux; lens; performance metrics; respirator; self contained breathing apparatus; SCBA; temperature; viewing section

NIST Special Publication 1123

Emergency First Responder Respirator Thermal Characteristics: Workshop Proceedings

Amy Mensch
Nelson Bryner

U.S. Department of Commerce
Engineering Laboratory
National Institute of Standards and
Technology
Gaithersburg, MD 20899

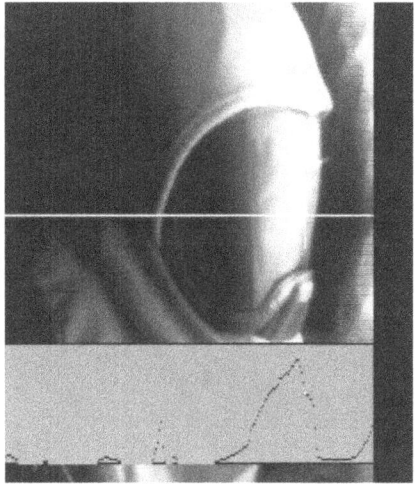

NIST National Institute of Standards and Technology • U.S. Department of Commerce

NIST Special Publication 1123

Emergency First Responder Respirator Thermal Characteristics: Workshop Proceedings

Amy Mensch
Nelson Bryner
U.S. Department of Commerce
Engineering Laboratory
National Institute of Standards and Technology
Gaithersburg, MD 20899

June 2011

The Fire Protection Research Foundation
Kathleen H. Almand, *Executive Director*

National Institute of Occupational Safety and Health
John Howard, *Director*

National Personal Protective Technology Laboratory
Les Boord, *Director*

Department of Homeland Security
Janet Napolitano, *Secretary*

Federal Emergency Management Association
W. Craig Fugate, *Administrator*

United States Fire Administration
Glenn A. Gaines, *Acting Administrator*

U.S. Department of Commerce
Gary Locke, *Secretary*

National Institute of Standards and Technology
Patrick D. Gallagher, *Director*

Certain commercial entities, equipment, or materials may be identified in this document in order to describe an experimental procedure or concept adequately. Such identification is not intended to imply recommendation or endorsement by the National Institute of Standards and Technology, nor is it intended to imply that the entities, materials, or equipment are necessarily the best available for the purpose.

National Institute of Standards and Technology Special Publication 1123
Natl. Inst. Stand. Technol. Spec. Publ. 1123, 52 pages (June 2011)

CODEN: NSPUE2

ABSTRACT

The purpose of this workshop was to identify performance needs and establish research priorities to address the thermal characteristics of respiratory equipment used by emergency first responders. The workshop provided a forum for representatives from the first responder community, self contained breathing apparatus (SCBA) and component manufacturers, and research and testing experts to discuss issues, technologies, and research associated with SCBA high temperature performance. The goals of the workshop were defined in two parts: 1) Clarify baseline information, including the current state-of-the-art, applicable fire service events, and current related research, and 2) Research planning, including identification of performance needs and short and long term research priorities.

Presentations were given to explain the current SCBA certification process, understand experience from actual fire service incidents, and review the current state of respirator research. After the presentations, the workshop divided into three working group sessions to discuss performance needs and research priorities in smaller groups. Suggested topics for discussion included: a) Current Equipment, b) Current Practice and Usage, c) Future Trends, d) Short Term Research Needs, e) Long Term Research Needs, and f) other issues.

The results of the three smaller groups' deliberations were discussed when the full workshop reconvened. The responses from each group were merged into a combination of issues that related to the use and performance of the lens of the SCBA. The primary concerns and research priorities were the characterization of the fire fighter environment, performance of current and new technology, development of representative and realistic testing, and improvements to fire fighter training on the limitations of protective equipment. A significant amount of discussion concentrated on the testing for NFPA certification, which currently contains limited thermal testing.

Keywords: fire fighting; first responder; heat flux; lens; performance metrics; respirator; self-contained breathing apparatus; SCBA; temperature; viewing section

ACKNOWLEDGEMENTS

The success of any workshop is dependent on the hard work of the individual speakers, facilitators, and participants. These proceedings are an assimilation of the contributions from everyone involved in the workshop; copies of the presentations are included in Appendix 3. Thanks to all who made presentations and to those who participated in the workshop.

Thanks go to Nelson Bryner from NIST, Casey Grant from FPRF, and Bill Haskell from NIOSH, who served as chairs of the breakout sessions and helped bring focus to the discussions. In addition, we wish to acknowledge the assistance of Les Boord, NPPTL director, who provided location and equipment for the workshop. The authors also appreciate Susan Buller from NIOSH and Wanda Duffin-Ricks from NIST for providing logistical support.

TABLE OF CONTENTS

ABSTRACT	iii
ACKNOWLEDGEMENTS	iv
INTRODUCTION	1
WORKSHOP ORGANIZATION AND OBJECTIVES	5
BREAKOUT GROUP RESULTS	8
CONCLUSIONS	12
REFERENCES	13
APPENDIX 1 – WORKSHOP AGENDA	15
APPENDIX 2 - WORKSHOP ATTENDEES	17
APPENDIX 3 – WORKSHOP PRESENTATIONS	19
APPENDIX 3.A – Workshop Purpose and Goals	19
APPENDIX 3.B – NIOSH/NPPTL Overview	22
APPENDIX 3.C – NFPA Certification of Fire and Emergency Services PPE	25
APPENDIX 3.D – Safety Equipment Institute Testing Laboratory	28
APPENDIX 3.E – NIOSH/FFFIPP Activities	30
APPENDIX 3.F – Massachusetts Fire Training Academy	37
APPENDIX 3.G – NIST Respirator Research	41
APPENDIX 4 – WORKING GROUP QUESTIONS	45
APPENDIX 5 – WORKING GROUP RAW RESULTS	47
APPENDIX 5.A – Working Group A Raw Results	47
APPENDIX 5.B – Working Group B Raw Results	49
APPENDIX 5.C – Working Group C Raw Results	51

EMERGENCY RESPONDER RESPIRATOR THERMAL CHARACTERISTICS

INTRODUCTION

First responders use a self-contained breathing apparatus (SCBA) in order to provide breathable air to the user in atmospheres that present immediate danger to life and health (IDLH). In the case of fire fighting, the environment often lacks oxygen and contains smoke, carbon monoxide and other toxic products of pyrolysis and combustion. The SCBA provides a clean supply of air and respiratory protection from these contaminants. In addition the SCBA provides some protection from heat. Since the personal protective ensemble covers the entire body, part of the SCBA is relied on to protect the face and respiratory tract from thermal injuries. A certain amount of thermal protection is provided by the thermal resistance of the materials themselves and by cooling from the air flow inside the SCBA.

The current SCBA is primarily comprised of three components: a high pressure tank, a pressure regulator system, and an inhalation connection. The inhalation connection in the typical design is a facepiece or mask, which seals around the entire face and chin. Straps are used to keep the facepiece secured on the face and maintain the seal. The low pressure regulator and hose assembly connects to the front of the facepiece, and when opened, supplies air to maintain a positive pressure (a pressure greater than ambient) inside the facepiece. A nosecup incorporated inside the facepiece directs the user's exhalation out of the facepiece through one way valves. This design helps reduce fogging of the lens, which would impair visibility. Polycarbonate is typically used for the lens material due to its superior impact resistance and clarity, and relatively good thermal properties. Often an abrasion resistance coating, such as silicon oxide, is applied to the polycarbonate to reduce scratching of the lens.

There are three main standards that exist for SCBA: 42 CFR Part 84, EN 136, and NFPA 1981. The National Institute of Occupational Safety and Health (NIOSH) is the authority responsible for testing and certifying respiratory equipment in the United States as documented in the Code of Federal Regulations, 42 CFR Part 84 – *Approval of Respiratory Protective Devices*.[1] SCBA is only one type of respirator that is covered in 42 CFR Part 84. The document specifies components and minimum requirements to certify respirators based on the effectiveness of respiratory protection provided in hazardous atmospheres. Testing involves evaluation of quantities such as device weight, impact resistance, service time, breathing resistance, gas flow, and inhalation and exhalation valve performance.

In Europe, the European Committee for Standardization (CEN) publishes the *European standard for respiratory protective devices – full face masks*, EN 136. This document specifies the requirements for approval of full face masks as part of respiratory protection devices. A full face mask is defined as "... a facepiece which covers the eyes, nose, mouth and chin and provides adequate sealing on the face of the wearer of a respiratory protective device against the ambient atmosphere, when the skin is dry or moist, and even when the head is moved or when the wearer is speaking."[2] Testing involves evaluation of temperature resistance, flammability, thermal radiation resistance, harness strength, speech diaphragm, visibility, inhalation and exhalation valve performance, leaktightness, breathing resistance, and inward leakage.

The National Fire Protection Association (NFPA) began publishing voluntary standards for respiratory equipment in May 1971, with NFPA 19B – *Standard on Respiratory Protective Equipment for Fire Fighters*. The 19B document prohibited filter-type canisters and only allowed SCBA. In May 1981, NFPA 19B was withdrawn, and NFPA 1981 – *Standard on Self-Contained Breathing Apparatus for Fire Fighters* was adopted, which required a positive pressure design and a minimum service time of 30 min. The NFPA standard required NIOSH certification, but additional requirements were necessary to capture the conditions specific to fire fighting. Performance requirements and appropriate testing protocols, which simulated environments experienced in fire fighting and storage were added in 1987. The first heat and flame exposure test was implemented in the 1992 edition. In 2002, a universal air connection was specified, so that any air source could replace an empty cylinder in an emergency situation. The most recent edition of NFPA 1981, adopted in 2007, was changed to *Standard on Open-Circuit Self-Contained Breathing Apparatus (SCBA) for Emergency Services*. Another significant change to the standard was the requirement for protection from chemical, biological, radiological and nuclear terrorism agents, or CBRN certification from NIOSH.[3] The NFPA 1981 document specifies minimum requirements for the NFPA certification for use by fire and emergency responders in atmospheres that are immediately dangerous to life and health (IDLH).

The thermal environments that fire fighters are exposed to in structural fires are highly variable, and depend on many factors including fuel type and load, interior finish, and structure layout and construction. Studies on fire fighter protective clothing [4-6] have described pre-flashover fire fighting environments with temperatures between 100 °C and 300 °C and maximum heat fluxes between 5 kW/m^2 and 12 kW/m^2. More dangerous fire fighting environments where protective clothing has been studied, involve temperatures up to 700 °C and heat fluxes of 20 kW/m^2 to 40 kW/m^2.[5-7] However, conditions of flashover and post-flashover can reach 1000 °C and 170 kW/m^2.[8] Donnelly et al. [9] combined various reports and articles from the literature, which classified fire fighting environments into categories and specified the maximum time, temperature and heat flux associated with each type of exposure. The result was a recommendation of four thermal classes of fire fighter exposure, which are displayed graphically in Figure 1. The maximum time for each class is listed within the shaded area showing the range of air temperatures and heat flux values at each thermal class. Thermal classes such as these can be used to establish performance requirements of protective equipment standards.

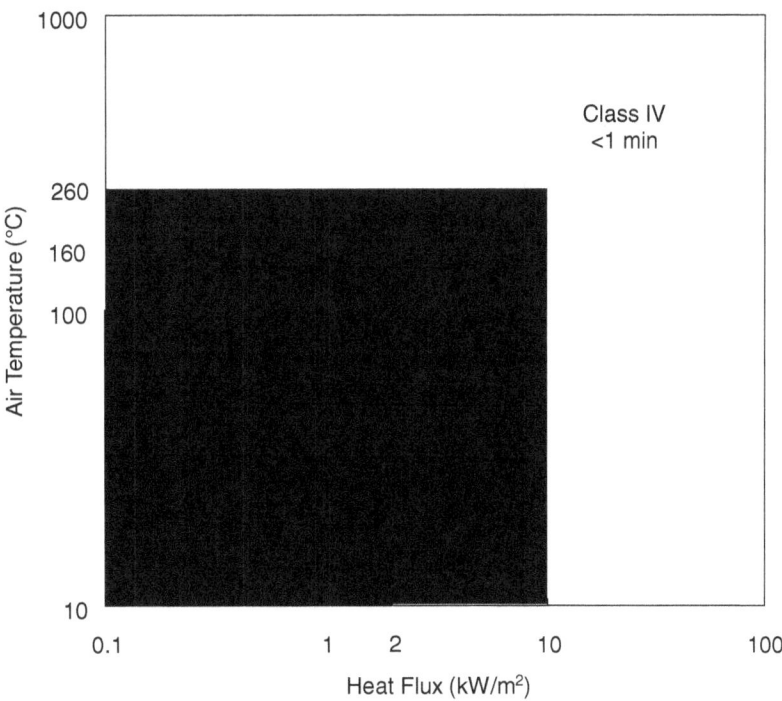

Figure 1 — Graphical representation of the recommendations for thermal classes of fire fighter environments, from Donnelly et al., 2006, [9] showing range of air temperature, heat flux and duration.

Currently in the US, the certification test that involves the most severe thermal exposure for an SCBA is the Heat and Flame Test, Section 8.11 of NFPA Standard 1981 — Open Circuit Self-Contained Breathing Apparatus (SCBA) For Emergency Services.[10] In this test, the SCBA is mounted on a test headform breathing at 40 L/min and placed in a convection oven at 95 °C for 15 min. This exposure would be classified as Class I in Figure 1. No more than 20 s later, the breathing rate is increased to 103 L/min, and the SCBA is exposed to direct flame contact for 10 s. The contact peak temperature is specified to be between 815 °C and 1150 °C. Therefore the second portion of the test fits into Class IV from Figure 1. Following the heat exposures, the headform is dropped from a height of 15.2 cm (6 in). The SCBA is tested for airflow performance and for visual acuity. Although this test involves elevated temperatures, it does not capture the conditions of temperature and heat flux that a fire fighter may experience in Classes II and III.

This workshop was designed to discuss the research needs in order to reduce the problem of heat related respirator failures during fire fighting. The need for improved SCBA and facepiece design to withstand a variety of extreme conditions including high heat loads, was documented in a U.S. Fire Administration special report in 2001.[11] In the decade since, several additional reports on fire fighter fatalities have indicated that inadequate thermal performance of the SCBA lenses contributed to one or more fire fighter fatalities.[12-18] The SCBA masks in most of these cases were found still on the victims, and all displayed extensive damage to the point where the SCBA could no longer provide respiratory protection from the IDLH environment. In addition, there have been numerous anecdotal accounts of crazing, bubbling, and softening of lenses, some of which have been reported as near misses.[19-22] These incidents have caught the attention of the first responder community and SCBA manufacturing industry. Because the SCBA plays such a critical role in the survival of fire fighters, especially those that find themselves trapped or lost, the SCBA user and manufacturing community has decided that the issue of respirator thermal performance needs to be addressed. This workshop was organized to identify the current state-of-the-art of existing SCBA technology, identify performance needs, and prioritize the research needs to reduce heat related respirator failures for fire fighters.

WORKSHOP ORGANIZATION AND OBJECTIVES

The primary objective of the workshop was to identify performance needs and establish research priorities to address the thermal characteristics of respiratory equipment used by emergency first responders. The workshop assembled participants from the first responder community, SCBA and component manufacturers, and research and testing experts to facilitate discussion from a variety of perspectives.

The issues, technologies, strategies, and research associated with the performance of respirators in high heat environments was presented and discussed. The workshop agenda and a list of attendees are provided in Appendices 1 and 2, respectively. The objectives of the workshop were defined in two parts:

> Clarify baseline information: including the current state-of-the-art, applicable fire service events, and current related research, and

> Research planning: including identification of performance needs and short and long term research priorities.

Presentations were given to clarify baseline information. The complete list of presentations and presenters can be found in the agenda in Appendix 1. The slides that were presented are provided in Appendix 3. Workshop participants were divided into smaller breakout groups, which identified performance needs and started the prioritization process. The breakout group results were presented to the workshop for review and discussion of research planning.

To describe the current self-contained breathing apparatus (SCBA) and certification process, representatives from the National Institute of Occupational Safety and Health (NIOSH)/National Personal Protective Technology Laboratory (NPPTL), the National Fire Protection Association (NFPA), and Safety Equipment Institute (SEI) each gave presentations. The presenters described the role of each of their organizations in the regulation, standards writing, and certification processes. It was explained that the NFPA 1981 committee on the open circuit SCBA was in the process of examining and accepting proposals for changes to the standard until the summer of 2011, to be implemented in 2013. The current standard [3] has a heat and flame test which was designed to eliminate melting and after-flaming of plastic components of the facepiece such as straps, hoses, and the speaking diaphragm, and to maintain visual acuity of the facepiece lens after a thermal impact in Class I and a short duration in Class IV from Figure 1. The committee has been considering adding another more severe thermal impact test, which would test the mechanical survivability of the facepiece lens.

To understand selected fire service events, representatives from the NIOSH Fire Fighter Fatality Investigation and Prevention Program (FFFIPP) described five incidents in recent years related to thermal degradation of respirators. There were seven fatalities associated with these events, in which the fire fighter's SCBA may have been thermally degraded while the user was still "on air." This was indicated by thermal injuries to the victims' respiratory systems, as well as the fact that the victims were found with their SCBA masks damaged, but still in place. It is assumed that if a fire fighter's SCBA tank runs out of air, the user's reaction would be to remove the mask; this was not observed in these cases. Although mask degradation may have contributed to these fatalities, there were also many other factors. A common aspect discussed was the occurrence of a rapid increase in thermal energy, and the attempt, but failure to escape. It is expected that conditions encountered exceeded the performance capabilities, and subsequently the NFPA requirements, of the SCBA. Three fire service organizations spoke about some of these events as well as their organization's overall experience related to the thermal issues with respirators. They expressed concern over the performance of the SCBA, as well as issues related to training and the loss of ability to sense the danger of the environment with the current use of highly insulating turnout gear and hoods. The users indicated two distinct needs regarding the SCBA and its relationship to the rest of the components of the personal protective ensemble:

> Further research to improve the thermal resistive performance and test methods.

> Increased awareness for the fire service community about the thermal protection levels provided.

To provide a review of the current state of respirator research, representatives from the Fire Protection Research Foundation (FPRF), NIOSH/NPPTL, and the National Institute of Standards and Technology (NIST) outlined related research developments and future plans. The FPRF and NPPTL reported that there were several recently completed and in-progress studies related to the fire service. The studies, which are related to respirator use and fit, other personal protective equipment, or fire fighting tactics, could reveal information applicable to the issue of damaged SCBA. Several potential research topics were proposed that related to the thermal characteristics of the SCBA or other equipment used by fire fighters, as well as re-evaluation of the fire environment for protective equipment design. NIST updated the participants on several ongoing projects related to respiratory issues or respirators. These included real-time particulate monitoring during overhaul, contaminant sensor location in respirators, respirator fit characterization, methods for thermal exposures, and respirator high temperature performance metrics. NIST has begun to characterize the thermal performance of respirators in lab scale radiant panel tests and oven tests, and in full scale fire experiments. Specific input from the workshop participants to help direct future research efforts was requested as experiments and processing of data continue.

At the conclusion of the presentations, which provided valuable background information, the workshop was divided into three working groups to discuss performance needs and research priorities. The groups were arranged to include a relatively balanced mix of users, manufacturers and researchers. The different participants brought unique and valuable perspectives to the discussion. The workshop organizers suggested several topics for discussion including:

    Current Equipment
    Current Practice and Usage
    Future Trends
    Short Term Research Needs
    Long Term Research Needs
    Other Issues

The detailed list of working group questions is located in Appendix 4. The groups were reminded to consider factors related to the changing fire service landscape. This included the effects of technological changes to the SCBA on the horizon (e.g. flat-pack design), enhancements to other protective equipment, the modern style of building construction and furnishings, and current tactics and training practices.

The first working group session focused on the performance needs and the second on research priorities. After each working group session, the workshop participants reconvened, and each working group reported a summary of their discussion, followed by an entire group discussion. The results of the sessions of the working groups were combined and are presented in the next section.

BREAKOUT GROUP RESULTS

The raw results of the breakout group discussions are located in Appendix 5. These results reflect items that were discussed during the sessions for *Identification of Performance Needs* and *Development of Research Priorities*. The items are a mixture of direct answers to working group questions, important related points or even questions and concerns raised through discussion within the breakout groups. The working groups typically discussed issues in both general and specific terms, which are reflected in the results. Although there was a wide variety of ideas and topics discussed, it was evident there were certain common themes. The same response was not always listed in exactly the same manner from all groups, but there was significant overlap. Conversely, identically listed responses may have had different meanings when taken within the context of the working group discussion.

As part of the discussion on performance needs, the working groups identified issues with the current equipment and practice as well as possible solutions. Thirteen items were mentioned by all three groups in the discussion of performance needs and are presented in Table 1, with seven items that identify issues, and six items that identify possible solutions. The items are listed as a general topic, and then more detail from each group's perspective is listed under "scope."

There were many items that were identified by only one or two of the groups, but this did not mean that these ideas do not have merit or that other groups didn't agree. The working group format certainly allows recognition of common themes, but it also relies on individuals to think of issues and ideas that may not be obvious to everyone else. These items can all be found in Appendix 5, but a few of the innovative items include reflective coatings, a rub/impact/wipe test, supplemental lenses, repeated exposures, temperature extremes, soot effects, active cooling, flat pack effects, and PASS issues.

As the discussion on research priorities followed, the groups began to focus their discussion on the most important issues, and how to address them. It became clear that many items were repeated from performance needs, but with more clarity. Table 2 lists four items that were mentioned by all three groups in the discussion of research priorities, in no particular order. The scope of each item is separated into aspects that relate to either short term or long term tasks. The research items that were listed by one or two groups include: thermal warning indicator, including other organizations in the research effort, establishment of a research clearinghouse, development of training media for the new generation of fire fighters, consistent ensemble testing, information overload for fire fighters, funding issues, repeated exposures, communications issues, and conducting a mask damage incident survey.

Table 1 – Common items for performance needs identified by all three groups.

Item	Scope	Type of Item
Facepiece is weakest link	Lens, mask, face piece, SCBA are weakest components.	Issue Identified
Variable environment	What is a representative exposure? Define the environment, but it's highly variable. That's why there is no such thing as routine fire fighting. All situations are important.	Issue Identified
Optimization Trade-offs	Optimize existing materials, performance, and cost, etc. Could be a trade-off for visibility and better thermal performance. ANSI-Z87.1 transparency test may not be passed by new materials.	Issue Identified
Training issues	Training concerns, insufficient live fire training, enforcement of training. Must train for personal responsibility. Create a multi-agency partnership on a training video.	Issue identified
Behavioral issues	Must address behavior and usage issues. Change the fire service culture.	Issue Identified
Care and maintenance	Look into the standard and Selection, Care and Maintenance (SCaM) documents. Need to better understand gear care and maintenance. Enforce visual inspection of equipment.	Issue Identified
PPE prevents sense of environment	The protection of the other PPE prevents a sense of hazard. Should we go backwards and reduce or limit protection of other PPE? Fire fighters had a better sense of the danger of the environment before the high levels of PPE were worn. Must improve mask performance, not lower other PPE, however, could mandate an upper limit for TPP (for hoods or gloves).	Issue Identified
Ensemble testing	Uniform testing of entire ensemble, for both severity and frequency. All elements should be equal in performance.	Possible Solution
Design for catastrophic event	Design so a mask failure is not fatal, to prevent catastrophic failure. Extend survivability time in emergency conditions. Design for a catastrophic event. Should survive one flashover.	Possible Solution
Alternative materials	Glass, composites, layering, new materials, polyethersulfone.	Possible Solution
Different uses	Different ratings for masks used in fire-fighting, overhaul, confined space, flashover training? Should use *same* masks for training and fire fighting.	Possible Solution
Warning Device/ indicator	Thermal warning device, imminent hazard indicator, pre-failure indicator, early warning method, built in warning system. Could be based on temperature, flux, or rate of temperature rise. Could use a worst case algorithm. Could have an audible alarm, straight forward signal, or a radiant heat or temperature indicator/ display, such as a bimetallic strip. Can use HUD.	Possible Solution
Reassess test methods	Reassess test methods, more realistic radiant exposure lens test. Radiant vs. convective exposure. Must be a repeatable sequence – might simulate rapid fire growth. Test parameters could go higher/lower depending on the need.	Possible Solution

Table 2 - Common items for research priorities identified by all three groups.

Item	Scope	Type of Item
Improve Mask Design	Identify mechanical/engineering solutions. Develop new materials, designs, approaches. Research new material properties − visual acuity, thermal, impact/abrasion resistance.	Short Term
	Implementation of an improved/enhanced mask material, design.	Long Term
Representative and realistic testing	Evaluate existing test methods. Test protocols should replicate "real world" conditions in reproducible test, and represent appropriate scenarios, such as a lost fire fighter or routine fire fighting.	Short Term
	Revise/develop new test methods. A new test method for heat exposure and/or heat and flame.	Long Term
Best Practices, Behavior/ Training	Adhere to best practices − involves training and behavior. Identify needs for changes to training. Identify solutions for changing behavior (human solutions). Address limitations for PPE in training. Educate approach for improvement.	Short Term
	Address training and enforcement needs and implement changes.	Long Term
Define Fire Environment	Identify thermal stresses (radiant/convective). Collect quantitative data on actual usage (flux, temperature, and time). Define heat flux/temperature levels for survivability. Define conditions during and after an extreme event.	Short Term
	Hold a "Project FIRES" part II, with national involvement. Measure real time data temperature on the respirator.	Long Term

A desire to have an improved facepiece design was an action item shared by all the groups. New materials, technologies and approaches were suggested. While polycarbonate has traditionally been used as the lens material in SCBA, a material with better thermal properties, such as polyether sulfone was suggested. Polyether sulfone has a heat deflection temperature at 0.46 MPa of about 210 °C as opposed to polycarbonate (about 140 °C).[23] The potential benefits and drawbacks of a hybrid glass and plastic lens were also discussed. Glass has superior thermal properties, and plastics have better impact resistance; however, differences in thermal expansion would be an issue for layered lenses, with the curved geometry that exists currently. The integration of a warning system, possibly with the heads-up-display (HUD), was suggested to indicate either facepiece temperatures at the limit of performance, or dangerous conditions (heat flux or temperatures) likely to damage a facepiece. Because of the improvements in the protection of the rest of the gear, it is a concern that fire fighters cannot sense how hot their environment actually is. Bubbling of the lens material often occurs when the lens softens and loses mechanical properties, but this usually cannot be seen because of dark, smoky conditions. Further, when the bubbling occurs, the danger already exists because the lens is in a softened state, susceptible to damage. More research and development is needed on new materials and designs before any or all of these changes could be implemented.

The second item calls for improvements to tests and standards. The latest NFPA 1981 Standard for Open Circuit SCBA For Emergency Services contains a Heat and Flame Test, Section 8.11.[3] In this test, the SCBA is mounted on a test headform, breathing at 40L/min, and placed in a convection oven at 95 °C for 15 min. No more than 20 s later, the breathing rate is increased to 103 L/min, and the SCBA is exposed to direct flame contact for 10 s, followed by a drop from a height of 15.2 cm (6 in). The SCBA is tested for airflow performance, but also for visual acuity. The participants of the workshop agreed that this does not capture the conditions of temperature and heat flux that a fire fighter may be exposed to in emergency situations. Also, considering that the rest of the fire fighter's personal protective equipment is tested in an oven at 260 °C, the SCBA has the lowest requirement in terms of thermal performance. Additional tests have been proposed by the NFPA committee that would test for "survivability" of the SCBA. The proposed tests would be similar to the Heat and Flame test, but certain parameters would be different, such as a higher oven temperature or a gas fired radiant heat panel exposure. The breathing rate would stay at 40 L/min for the entire test. The only pass criteria would be for the SCBA to maintain positive pressure for the duration of test. The NFPA committee is open to feedback regarding this test or other ideas. In general, there were several concerns raised with regard to testing. It was agreed that the tests should replicate realistic conditions as much as possible, while being repeatable and manageable in cost. Laboratory ovens and small flames are repeatable, but may not be representative of real conditions. The current test also may be missing the component of radiative heat transfer, which can be dominant over convective heat transfer under certain fire fighting conditions. Many participants believed that the entire personal protective equipment ensemble should be tested together. This has not been done in the past, because of the logistical issues of bringing together equipment from various manufacturers. However, ensemble testing would ensure that all components were tested at the same level.

Changes to the next edition of the NFPA 1981 standard will not go in effect until the next revision, tentatively set for 2013. In addition, a new standard only applies to new equipment. Departments typically replace equipment at regular intervals, but considering these lag times, it would take up to 15 years to change out all of the equipment. This is one reason why many participants stressed that changes to training would have a more immediate impact than changes to equipment. A combined effort among multiple agencies was suggested to develop a video demonstrating the limitations of SCBA equipment and the dangers associated with certain behaviors. It was expressed that effective training is important, no matter what improvements are made to the gear. There will always be some limitations to the gear, and fire fighters need to be aware of the limitations.

Understanding the thermal stresses (temperatures, heat fluxes, and duration) associated with the fire fighting environment is critical to understanding the performance needed for SCBA. Studies with this kind of data are published regularly, but information has not been organized. In order to be useful and accessible, the specific data should be combined into general categories of exposures and reported in a single document. Additionally, studies measuring temperatures and heat fluxes on the SCBA and lens during actual use were suggested. Another idea for gathering "real world" information was to develop a survey for fire fighters on the prevalence of SCBA thermal damage and the conditions experienced by the mask.

CONCLUSIONS

A diverse group of experts in SCBA manufacturing, certification testing, use, and research was assembled to discuss and identify needs and research priorities related to emergency responder respirators. The first goal of the workshop was to review the current state-of-the-art of SCBA technology and identify critical performance needs. A summary of the performance needs defined by the workshop includes:

- Thermal resistance of the facepiece equal to or better than the rest of the fire fighter ensemble
- Understanding of and ability to sense the variable fire environment
- Responsible fire fighting culture, behavior and usage of SCBA.

After the performance needs were discussed, four areas were identified as priorities for further research and effort. The group presented both short and long term thrusts for each research area, including to:

- Improve SCBA facepiece design
- Characterize the fire environment
- Develop more representative and realistic testing
- Define best practices for use and behavior, and implement into training.

REFERENCES

[1] CFR, "Approval of Respiratory Protective Devices", Code of Federal Regulations Title 42 Part 84, 498-557, 2009.

[2] EN 136 Respiratory protective devices - Full Face masks - Requirements, testing, marking, 1998 Edition. European Committee for Standardization (CEN), rue de Stassart 36, B-1050 Brussels, Belgium.

[3] NFPA 1981 Standard on Open-Circuit Self-Contained Breathing Apparatus (SCBA) for Emergency Services, 2007 Edition. National Fire Protection Association (NFPA), Quincy, MA.

[4] Krasny, J. F., J. A. Rockett, and D. Huang, "Protecting Fire Fighters Exposed in Room Fires: Comparison of Results of Bench Scale Test for Thermal Protection and Conditions During Room Flashover", *Fire Technology 24* (1), 1988, pp 5-19.

[5] Lawson, J. R., "Fire Fighter's Protective Clothing and Thermal Environments of Structural Fire Fighting," NISTIR 5804, 1996, National Institute of Standards and Technology, Gaithersburg, MD.

[6] Rossi, R., "Fire Fighting and its Influence on the Body", *Ergonomics 46* (10), 2003, pp 1017-1033.

[7] Schoppee, M. M., J. M. Welsford, and N. J. Abbott, "Protection Offered by Lightweight Clothing Materials to the Heat of a Fire," ASTM STP 900, 1986, American Society for Testing and Materials, Philadelphia, PA.

[8] Fang, J. B and J. N. Breese, "Fire Development in Residential Basement Rooms," NBSIR 80-2120, 1980, National Bureau of Standards (currently NIST), Gaithersburg, MD.

[9] Donnelly, M. K., W. D. Davis, J. R. Lawson, and M. J. Selepak, "Thermal Environment for Electronic Equipment Used by First Responders," NIST TN 1474, 2006, National Institute of Standards and Technology, Gaithersburg, MD.

[10] NFPA 1981 Standard on Open-Circuit Self-Contained Breathing Apparatus (SCBA) for Emergency Services, 2007 Edition. NFPA, Quincy, MA.

[11] Thiel, A. K., "Special Report: Prevention of Self-Contained Breathing Apparatus Failures," USFA-TR-088, 2001, U.S. Fire Administration, Emmitsburg, MD.

[12] Madrzykowski, D., "Fatal Training Fires: Fire Analysis for the Fire Service, Interflam 2007," International Interflam Conference, 11th Proceedings, September 3, 2007, London, England, pp 1169-1180.

[13] NIOSH, "Career officer injured during a live fire evolution at a training academy dies two days later - Pennsylvania," Report F2005-31, 9-19-2007, National Institute of Occupational

Health and Safety, Fire Fighter Fatality Investigation and Prevention Program, Morgantown, WV, Available: http://www.cdc.gov/niosh/fire/reports/face200531.html .

[14] NIOSH, "Career Lieutenant and Fire Fighter Die in a Flashover During a Live-Fire Training Evolution - Florida," Report F2002-34, 6-16-2003, National Institute of Occupational Health and Safety, Fire Fighter Fatality Investigation and Prevention Program, Morgantown, WV, Available: http://www.cdc.gov/niosh/fire/pdfs/face200234.pdf.

[15] NIOSH, "Career Fire Fighter Dies in Wind Driven Residential Structure Fire - Virginia," Report F2007-12, 6-10-2008, National Institute of Occupational Health and Safety, Fire Fighter Fatality Investigation and Prevention Program, Morgantown, WV, Available: http://www.cdc.gov/niosh/fire/reports/face200712.html .

[16] NIOSH, "A Volunteer Mutual Aid Captain and Fire Fighter Die in a Remodeled Residential Structure Fire - Texas," Report F2007-29, 11-3-2008, National Institute of Occupational Health and Safety, Fire Fighter Fatality Investigation and Prevention Program, Morgantown, WV, Available: http://www.cdc.gov/niosh/fire/pdfs/face200729.pdf.

[17] NIOSH, "Career Probationary Fire Fighter and Captain Die as a Result of Rapid Fire Progression in a Wind-Driven Residential Structure Fire - Texas," Report F2009-11, 4-8-2010, National Institute of Occupational Health and Safety, Fire Fighter Fatality Investigation and Prevention Program, Morgantown, WV, Available: http://www.cdc.gov/niosh/fire/pdfs/face200911.pdf.

[18] NIOSH, "Volunteer Fire Fighter Dies While Lost in Residential Structure Fire - Alabama," Report F2008-34, 6-11-2009, National Institute of Occupational Safety and Health, Fire Fighter Fatality Investigation and Prevention Program, Morgantown, WV, Available: http://www.cdc.gov/niosh/fire/pdfs/face200834.pdf.

[19] National Fire Fighter Near Miss Reporting System, "Firefighter experiences near miss in flashover trailer training.," Report 06-441, 8-23-2006, Available: http://www.firefighternearmiss.com.

[20] National Fire Fighter Near Miss Reporting System, "Engine crew surprised by sofa flare up.," Report 06-428, 8-18-2006, Available: http://www.firefighternearmiss.com.

[21] National Fire Fighter Near Miss Reporting System, "Facepiece damaged during live burn training.," Report 07-903, 5-7-2007, Available: http://www.firefighternearmiss.com.

[22] National Fire Fighter Near Miss Reporting System, "Problem with CAFS unit identified at live burn.," Report 08-044, 1-25-2008, Available: http://www.firefighternearmiss.com.

[23] MatWeb Material Property Data. 2010 [cited 8-11-2010]; Available from: URL:http://www.matweb.com

APPENDIX 1 – WORKSHOP AGENDA

Agenda: Workshop on Emergency First Responder Respirator Thermal Characteristics

Time	Description	Who
	Day One - Tuesday, 27 July 2010	
9:00	Welcome, Preliminaries, and Introductions	*Casey Grant, Fire Protection Research Foundation*
9:20	Workshop Purpose and Goals and FPRF Research Update	*Casey Grant, Fire Protection Research Foundation*
9:30	<u>Status Review of Current State-of-the-Art</u>	
	NIOSH/NPPTL Overview	*Les Boord, National Personal Protective Technology Laboratory*
	NFPA Certification of Fire and Emergency Services PPE	*Bruce Teele, National Fire Protection Association*
	NFPA Technical Committee on Respiratory Protection	*Dan Rossos, Portland Fire and Rescue*
	Safety Equipment Institute Testing Laboratory	*Pat Gleason and Steve Sanders, Safety Equipment Institute*
10:00	Break	
10:10	<u>Review & Discussion of Applicable Fire Service Events</u>	
	NIOSH/FFFIPP Activities	*Stephen Miles and Tim Merinar, Fire Fighter Fatality Investigation and Prevention Program*
	Houston Fire Department	*Carl Matjeka, Houston Fire Department*
	Pennsylvania Fire Training Academy	*Edward Mann and Pat Pauly, PA State Fire Commissioner*
	Massachusetts Fire Training Academy	*Fred LeBlanc, Massachusetts Fire Training Academy*
11:00	<u>Review & Discussion of Current Related Research Initiatives</u>	
	NIST Respirator Research	*Nelson Bryner, National Institute of Standards and Technology*
	NIOSH/NPPTL Research	*Heinz Ahlers, National Personal Protective Technology Laboratory*
12:00	Lunch	

Time	Description	Who
1:00	<u>Identification of Performance Needs</u> Discuss Working Group Questions, side 1)	*Three breakout groups, A, B, C*
3:15	Break	
3:30	Presentation of Performance Needs from Breakout Groups	*Three breakout groups, A, B, C*
4:00	Group Discussion of Performance Needs	
5:00	Adjourn for the day	

Day Two - Wednesday, 28 July 2010

Time	Description	Who
8:00	<u>Development of Research Priorities</u> Discuss Working Group Questions, side 2)	*Three breakout groups, A, B, C*
9:30	Break	
9:45	Presentation of Research Priorities from Break-out Groups	*Three breakout groups, A, B, C*
10:15	Group Discussion of Research Priorities	
11:00	Break	
11:15	Summary	
12:00	Adjournment	

APPENDIX 2 - WORKSHOP ATTENDEES

SCBA Thermal Characteristics Workshop 27-28 July 2010 Pittsburgh, PA

	First Name	Last Name	Organization	Email Address
1	Heinz	Ahlers	NIOSH-NPPTL	hahlers@cdc.gov
3	Chris	Anaya	Sacramento Metro FD	anaya@prodigy.net
4	Eric	Beck	MSA Company	eric.beck@msanet.com
5	David	Bernzweig	Columbus Div of Fire	vpbernzweig@local67.com
6	Mark	Black	Naval Surface Weapons Ctr.	marshall.black@navy.mil
7	Les	Boord	NIOSH-NPPTL	zfx2@cdc.gov
9	Nelson	Bryner	NIST	nelson.bryner@nist.gov
11	Rich	Duffy	IAFF	rduffy@iaff.org
12	William	Flint	DC Fire & EMS Dept.	william.flint@dc.gov
13	Pat	Gleason	Safety Equip Institute	pgleason@seinet.org
14	Casey	Grant	FPRF	cgrant@nfpa.org
15	Ira	Harkness	Naval Surface Weapons Ctr.	a.harkness@navy.mil
16	Bill	Haskell	NIOSH-NPPTL	czj8@cdc.gov
17	John	Kuhn	MSA Company	john.kuhn@msanet.com
18	Fred	LeBlanc	MA Fire Academy	Redknight44@verizon.net
19	Jim	LeBlanc	Fosta Tek Optics	jleblanc@fosta-tek.com
20	Nick	Luzie	Sperian Resp Protection	nluzie@sperian.com
21	Edward	Mann	PA State Fire Commissioner	emann@state.pa.us
22	Craig	Martin	Avon-ISI	craig.martin@avon-rubber.com
23	Carl	Matjeka	Houston Fire Dept.	Carl.Matejka@houstontx.gov
24	Amy	Mensch	NIST	amy.mensch@nist.gov
25	Tim	Merinar	NIOSH-FFFIPP	tmerinar@cdc.gov
26	Stephen	Miles	NIOSH-FFFIPP	smiles@cdc.gov
27	Paul	Moore	NIOSH-FFFIPP	phm0@cdc.gov
28	Pat	Pauly	PA State Fire Commissioner	ppauly@state.pa.us
29	Jeff	Peterson	NIOSH-NPPTL	jpeterson1@cdc.gov
30	Jerry	Phifer	Tyco/Scott	jphifer@tycoint.com
31	Amy	Quiring	Tyco/Scott	astaubs@tycoint.com
32	Stephen	Raynis	FDNY	rayniss@fdny.nyc.gov
33	Daniel	Rossos	Portland Fire & Rescue	dan.rossos@portlandoregon.gov
34	Steve	Sanders	Safety Equipment Institute	ssanders@seinet.org
35	Robert	Sell	Dräger Safety, Inc	robert.sell@draeger.com
36	Angie	Shepherd	NIOSH-NPPTL	dlq0@cdc.gov
37	Michael	Shrum	Houston Fire Dept.	michael.shrum@houstontx.gov
38	Denise	Smith	Skidmore College	dsmith@skidmore.edu
39	Jon	Szalajda	NIOSH-NPPTL	zfx1@cdc.gov
40	Bruce	Teele	NFPA	bteele@nfpa.org
41	Bob	Timko	NVFC (PA Rep)	btimko@msn.com

42	Bill	Troup	FEMA	bill.troup@dhs.gov
43	Bruce	Varner	Santa Rosa FD	bvarner@santarosafd.com
44	Steve	Weinstein	Sperian Resp Protection	sweinstein@sperian.com
45	John	Williams	NIOSH-NPPTL	wjwilliams@cdc.gov

APPENDIX 3 – WORKSHOP PRESENTATIONS

APPENDIX 3.A – Workshop Purpose and Goals
Casey Grant, Fire Protection Research Foundation

Fire Protection Research Foundation
Update of Current Research
A) Recently Completed Studies Relating to Fire Service
- *Fire Ground Tactics for Alternative Energy Applications*
- Reaching the U.S. Fire Service with Hydrogen Safety Information
- *Fire Fighting Tactics Under Wind Driven Conditions*
- *Measuring Code Compliance Effectiveness*
- *Thermal Capacity of Fire Fighter Protective Clothing*
- Hazardous Materials Codes and Threshold Quantity
- *Respiratory Exposure Study*

(Completed Reports available at: www.nfpa.org/foundation)

Fire Protection Research Foundation
Update of Current Research
B1) On-going Projects (Administered by FPRF):
- Quantitative Evaluation of Fire & EMS Mobilization Times
- Developing Friction Loss Coefficients for Modern Fire Hose
- Fire Fighter Training for Advanced Electric Vehicles
- Hazard Assessment of Fire Service Training Fires

Fire Protection Research Foundation
Update of Current Research
B2) On-going Projects (with FPRF Advisory Services):
- Risk Factors for Fire Fighter Cardiovascular Disease (UArizona)
- Whole Glove Testing Technologies to Advance Perf. Standards for Structural Firefighting Gloves (NCSU)
- Evaluation of Stair Descent Devices (U of Illinois, Chicago)
- *Fireground Injuries: An International Eval. of Causes and Best Practices (UArizona)*
  - Columbus OH, Lancashire, Melbourne, Phoenix AZ, Prince William VA, Stayton OR, Tokyo (& Izumo, Matsue, Kiakyushu, Beppu, Osaka), Toronto, Wash DC
  - Others? (especially Hong Kong or European Fire Dept's)

Fire Protection Research Foundation
Update of Current Research
C) Potential Topics (relating to respiratory equipment):
- Evaluation of SCBA Face-piece Thermal Characteristics
- Field Usage of Deployed PPE Equipment
- Re-evaluation of the Fire Environment for PPE Design
- Performance Requirements for Compatible and Interoperable Electronic Equipment for Emergency First Responders
- PASS Alert Signal Analysis and actuation sequence

Fire Protection Research Foundation
Example of Changes on the Horizon
- New pressurized vessel technology (Flatpack)
- What is needed for infrastructure support?
  - Operability
  - Maintenance
  - Training / Behavioral

RESPIRATOR THERMAL CHARACTERISTICS WORKSHOP

AGENDA

1) Current State-of-the-Art

2) Applicable Fire Service Events

3) Current Related Research Initiatives

4) **Break-Out Groups:**
   - Performance Needs
   - Research Priorities

Contact Information:

Casey Grant

Fire Protection Research Foundation

One Batterymarch Park, Quincy, MA USA 02169-7471

Phone: 617-984-7284   Email: cgrant@nfpa.org

FPRF Website: www.nfpa.org/foundation

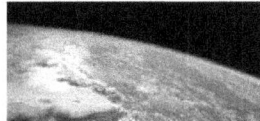

APPENDIX 3.B – NIOSH/NPPTL Overview
Les Boord, National Personal Protective Technology Laboratory

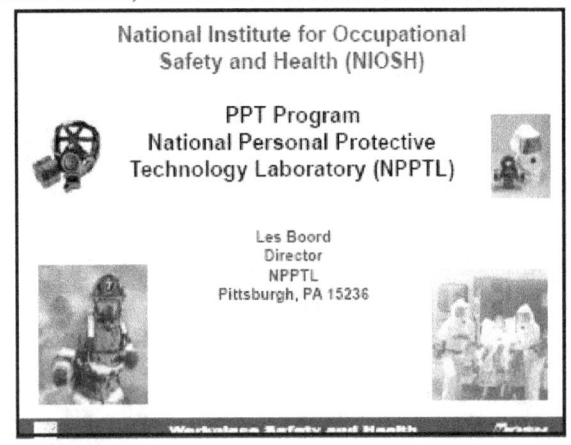
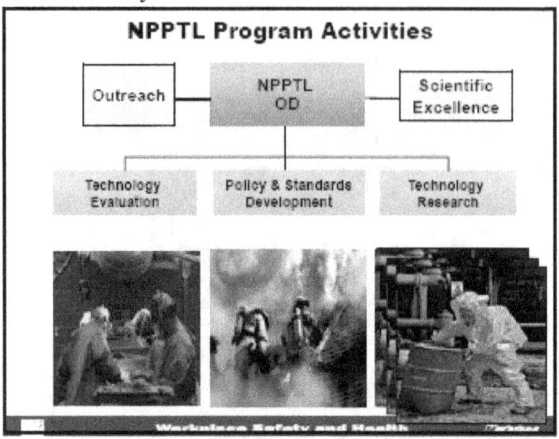
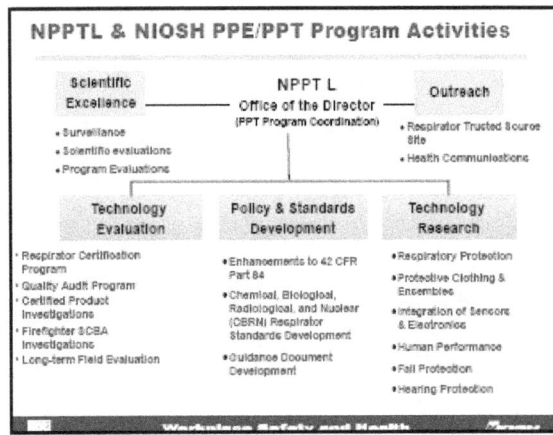

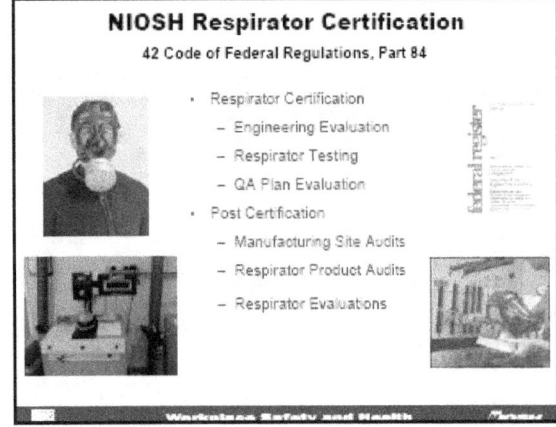
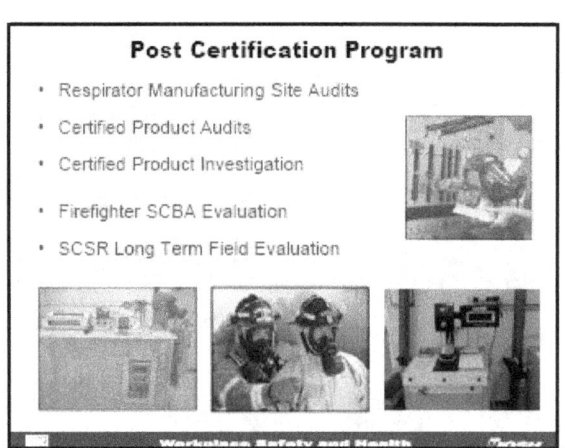

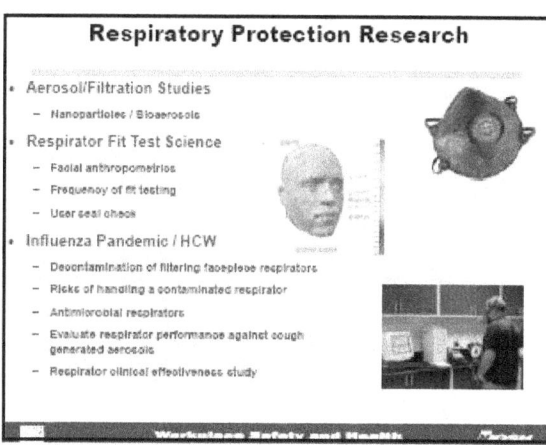
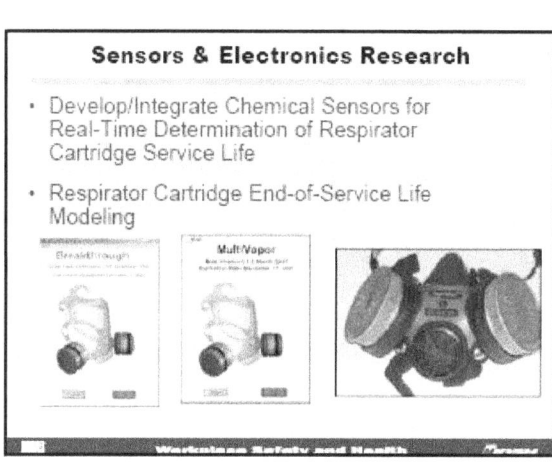
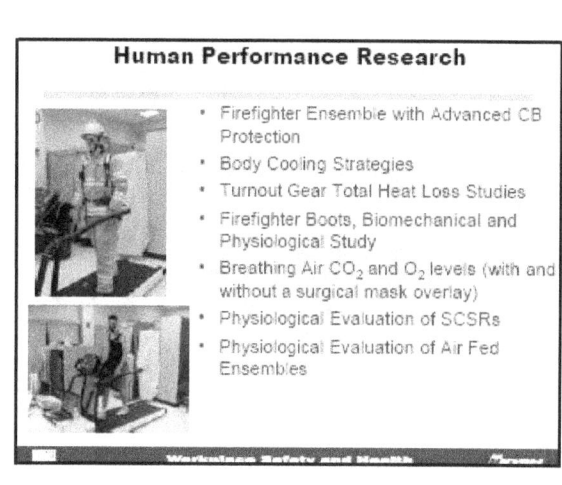
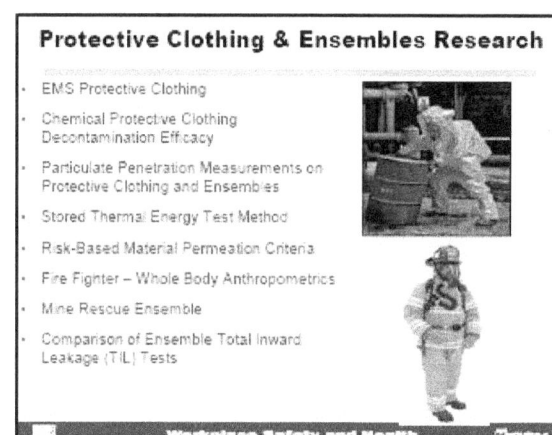
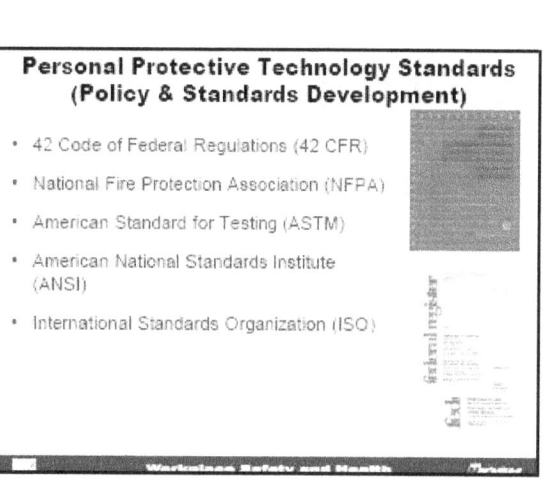

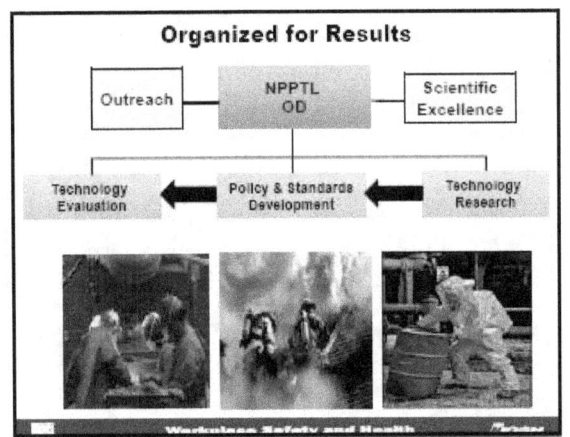
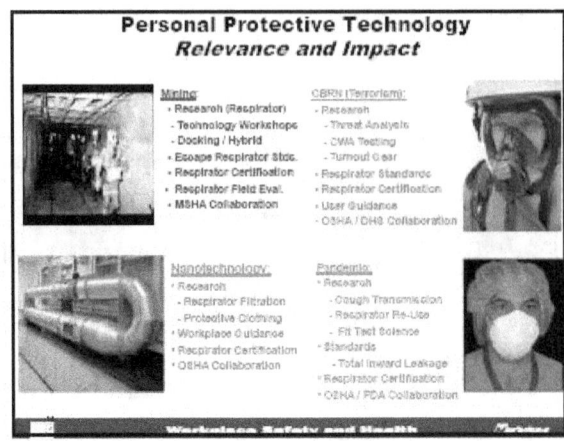
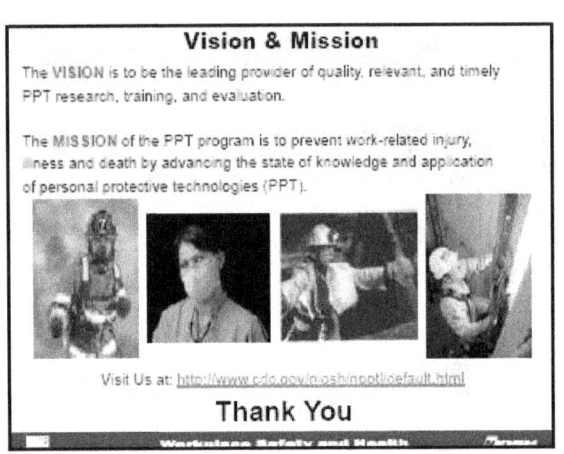

APPENDIX 3.C – NFPA Certification of Fire and Emergency Services PPE

Bruce Teele, National Fire Protection Association

CERTIFICATION OF FIRE AND EMERGENCY SERVICES PPE

DO EMERGENCY SERVICES PPE PRODUCTS COMPLY WITH STANDARDS??

- *NFPA DOES NOT EVALUATE, TEST, OR APPROVE ANY PRODUCTS OR SYSTEMS.*
- ALL NFPA standards for emergency services PPE require that products be *CERTIFIED*.
- Each NFPA PPE "product" standard contains the requirements for certification to that standard in Chapter 4.

CERTIFICATION =

- Cert Org: independent 3d party accredited to ISO 65.
- Cert org labs accredited for PPE to ISO 17025.
- Accreditation body operates under ISO 17011.
- Cert org performs inspection, evaluation, and testing to NFPA standard to determine product compliance.
- Compliant product bears <u>cert org's mark, label, and statement of compliance with NFPA standard</u> (with specific wording).

CERTIFICATION =

- Product *labeled* and *listed* by certification org.
- Mfg quality assurance program in accord with ISO 9001; *product always evaluated to NFPA standard*.
- Follow-up program at least twice each year.
- Annual verification of product compliance.
- Certification program also requires manufacturer investigation of complaints, product hazards, and have programs for safety alerts, product recall.

THIRD PARTY CERTIFICATION IS YOUR BEST ASSURANCE OF GETTING PRODUCT THAT IS ACTUALLY COMPLIANT

DON'T FALL FOR...

It will be certified soon.
Certification labels are expensive. I'll sell you the same thing without the label and save you lots of money!
It passes all the NFPA tests, I wouldn't lie to you.
It meets the intent of the standard.
This model is just as good as the certified ones.
Believe me, I wouldn't sell it to you if it wasn't safe.

Slide 1
To your question,
"Where are the certification labels?"

DON'T FALL FOR...

This is a display item; we don't put labels on them.
or
We were in a hurry to get this to you and I guess we forgot the label.
or
Don't worry, it's certified, you can take my word on it.

OR ANY OTHER EXCUSE!

Slide 2
ALWAYS THE PURCHASER'S *RIGHT*
TO REQUIRE SUBSTANTIATION OF CERTIFICATION

SPECIFY CERTIFICATION IN THE REQUESTS FOR BIDS AND IN THE PURCHASE SPECIFICATIONS

Slide 3
ALWAYS THE MANUFACTURER'S *RESPONSIBILITY*
TO PROVIDE SUBSTANTIATION OF CERTIFICATION BY SPECIFIC MODEL, TYPE, TRADE NAME OF PRODUCT.

PURCHASERS! *ALWAYS* VERIFY THE VALIDITY OF CERTIFICATION WITH THE CERTIFICATION ORGANIZATION *BEFORE SIGNING CONTRACT* !

Slide 4
NO CERTIFICATION ORGANIZATION LABEL WITH LOGO AND MARK ? ?

IT ISN'T CERTIFIED!

REGARDLESS OF WHAT THE MANUFACTURER OR DEALER CLAIMS !

Slide 5
PROJECT ON FIRE AND EMERGENCY SERVICES PROTECTIVE CLOTHING AND EQUIPMENT

Established in January 1995

The Project, consisting of the TCC and 7 Technical Committees, is responsible for product standards for protective clothing and equipment for fire and other emergency services personnel, and for standards on selecting, purchasing, maintaining, and retiring protective clothing and equipment.

Slide 6
PROJECT ON FIRE AND EMERGENCY SERVICES PROTECTIVE CLOTHING AND EQUIPMENT

TECHNICAL CORRELATING COMMITTEE (TCC)

- **Responsible for Project management** including document scopes, certification, and definitions; consistency of test methods for same exposures and hazards, and consistency of requirements for design, performance, test methods, certification, and protective clothing and equipment selection, care, and maintenance.

Chairman Les Boord
Director, NIOSH NPPTL

PROJECT ON FIRE AND EMERGENCY SERVICES PROTECTIVE CLOTHING AND EQUIPMENT

The 7 TECHNICAL COMMITTEES

- TC on Electronic Safety Equipment
- TC on Emergency Medical Services Prot Clothing & Equip
- TC on Hazardous Materials Protective Clothing & Equip
- TC on Respiratory Protection Equipment
- TC on Special Operations Protective Clothing & Equipment
- TC on Structural and Proximity Fire Fighting Protective Clothing & Equipment
- TC on Wildland Fire Fighting Protective Cloth & Equipment

EMERGENCY SERVICES PPE STANDARDS

NFPA 1801, Thermal Imagers for the Fire Service
NFPA 1951, Prot Ensembles for Technical Rescue Incidents
NFPA 1952, Surface Water Operations Prot Clothing and Equip
NFPA 1971, Prot Ensembles for Structural FF and Proximity FF
NFPA 1975, Station/Work Uniforms
NFPA 1977, Wildland FF Protective Clothing and Equipment
NFPA 1981, Open-Circuit SCBA for Emergency Services
NFPA 1982, Personal Alert Safety Systems (PASS)
NFPA 1983, Life Safety Rope and Equipment
NFPA 1984, Respiratory Prot Equip for Wildland FF Operations
NFPA 1989, Breathing Air Quality for Emergency Services
...continued

EMERGENCY SERVICES PPE STANDARDS, continued

NFPA 1991, Vapor Protective Ensembles for Haz Mats
NFPA 1992, Liquid-Splash Protective Ensembles for Haz Mats
NFPA 1994, Protective Ensembles for CBRN Terrorism Incidents
NFPA 1999, Emergency Medical Protective Clothing & Equip
NFPA 1851, Select, Care, Maint for Structural & Proximity PPE
NFPA 1852, Select, Care, Maint for Open-Circuit SCBA

APPENDIX 3.D – Safety Equipment Institute Testing Laboratory
Pat Gleason and Steve Sanders, Safety Equipment Institute

Certification of Fire and Emergency Services SCBA

NIOSH, NIST, NFPA/FPRF
Workshop on Emergency First Responder Respirator Thermal Characteristics
July 27, 2010

Certification of Personal Protective Equipment

Who is SEI?

- Non profit established in 1981 to administer the first non-governmental, third-party certification program to test and certify safety and protective equipment
- Voluntary program open to all manufacturers
- Policies developed by SEI Board of Directors which include representatives from: organized labor, industrial users, insurance industry, fire service users, and a manufacturer
- Accredited as Certification Body to ISO Guide 65 by ANSI & SCC

Accreditation Requirements

- ISO/IEC Guide 65 General Requirements for bodies operating product certification Systems
- References:
  - ISO/IEC guides related to accreditation and product certification
  - ISO/IEC 17025
  - ISO/IEC 17020
  - Nationally Recognized Product standards, NFPA, etc.

Certification of Personal Protective Equipment

SEI Mission

- assist government agencies, users, and manufacturers in meeting their mutual goals of protecting those who use safety and protective equipment on or off the job.

- aid advancements in protective equipment technology, using recognized standards and state-of-the art test facilities.

- support protective equipment users by providing an easily recognized mark for certified products.

Certification of Personal Protective Equipment

Meeting Industry Needs

- To fill void, certification programs in 1981 were initiated with seed money from National Safety Council, corporate users of PPE, and manufacturers

- Programs grew from head protection, eye & face protection, fire fighter helmets, gas detector tube units to all types of fire & emergency services PPT

Certification Programs Initiated

Compelling Factors

- Universal interest but no Federal mandate for certification of PPE beyond respirators
- Poor quality products being sold in North America
- False claims of compliance and certification
- Anxiety among users
  - highly dangerous occupations
  - injury rates reported by NSC, BLS
- Concerned manufacturers

Certification Process for Manufacturer

- Application for product certification including submission of documentation, drawings, materials & components
- Signs contract agreeing to terms of certification
- Product samples submitted for testing in accordance with performance standard at ISO/IEC 17025 accredited laboratory
- Review of design requirements, labels and user information
- On site quality assurance audit of entire Manufacturing Process
- Confirmation of appropriate Liability Insurance
- Surveillance includes annual recertification testing and quality assurance audits
- Registration to ISO 9001

SCBA Certification Program

Program Initiated in 1992, certifying to NFPA 1981-1992

Intertek – SEI's contract testing laboratory, built an SCBA test laboratory to support the program

At that time, NFPA 1981 only required NIOSH 42 CFR Part 84 approval for SCBA

In late 2001, NIOSH announced its SCBA CBRN approval program. This program required NFPA 1981 certification as a prerequisite.

SCBA Certification Program

In 2002, NIOSH and SEI developed a cooperative program whereby SEI would confirm NFPA 1981 certification of SCBA for NIOSH CBRN approval

Then in 2007, NFPA 1981 was revised to require NIOSH CBRN approval as a condition of certification

SCBA Certification Program

In 2007, SEI and NIOSH signed a Memorandum of Understanding (MOU)

As part of the MOU, SEI and NIOSH agreed to work cooperatively with regard to technical information, testing, certifications/approvals, standards development, investigations and attendance at public meetings

SEI and NIOSH staff work closely to support the MOU

Certification of Personal Protective Equipment

Pat Gleason
Steve Sanders
Safety Equipment Institute
703/442-5732
pgleason@seinet.org
www.SEInet.org

APPENDIX 3.E – NIOSH/FFFIPP Activities
Stephen Miles and Tim Merinar, Fire Fighter Fatality Investigation and Prevention Program

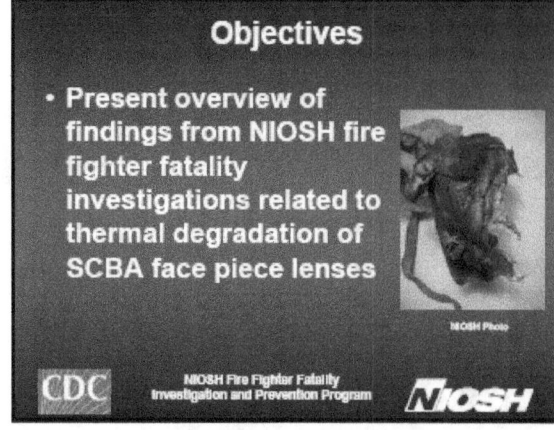

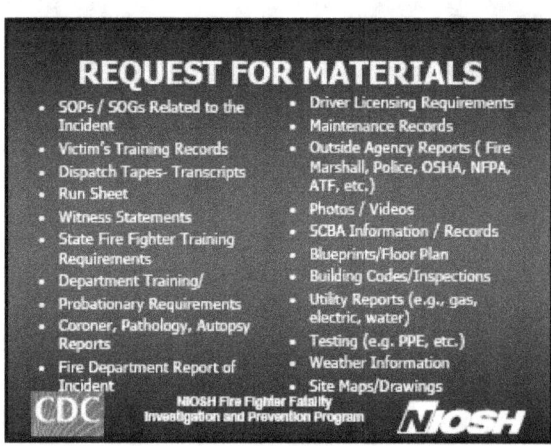
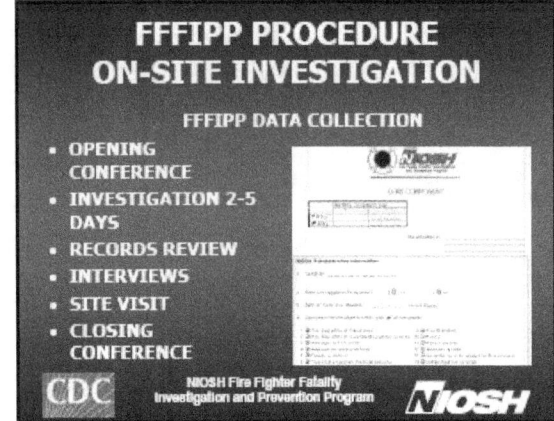
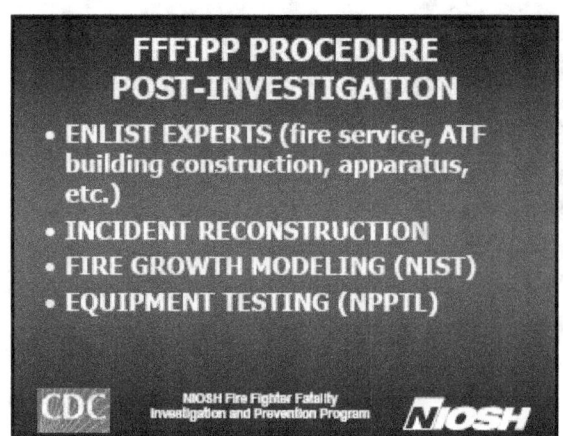

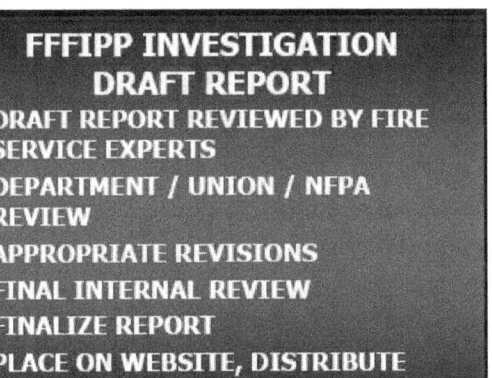
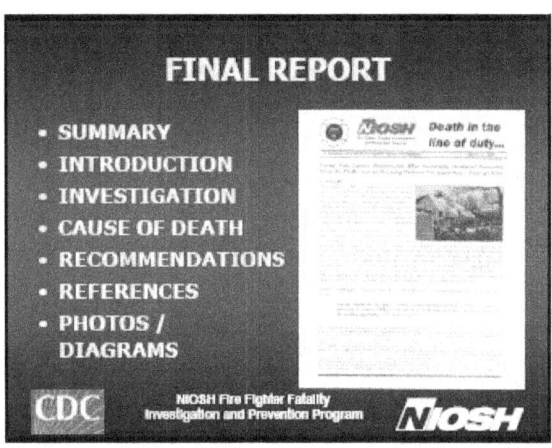
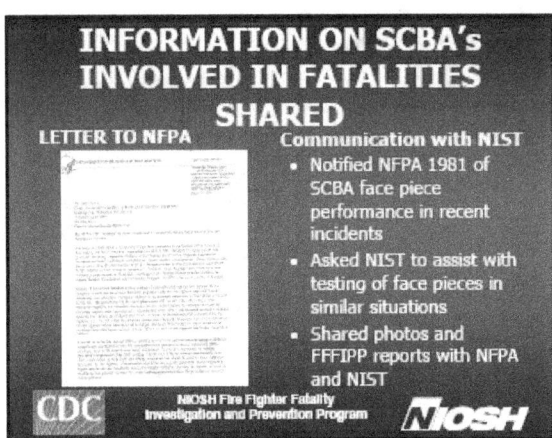
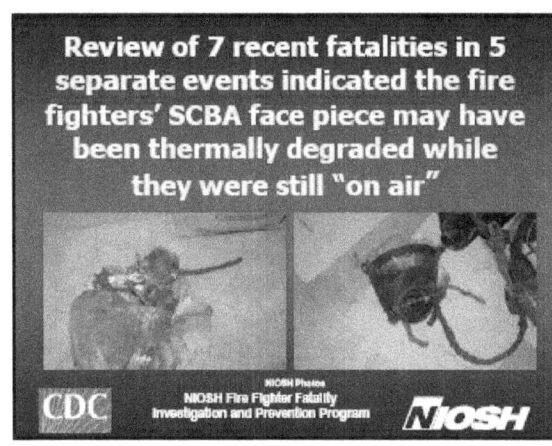
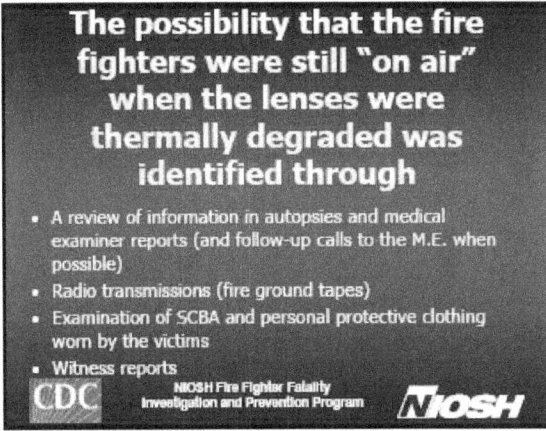
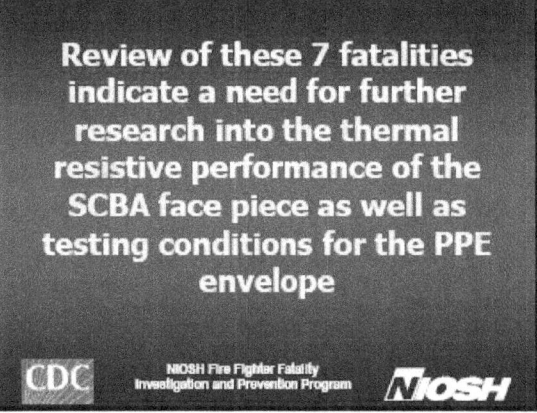

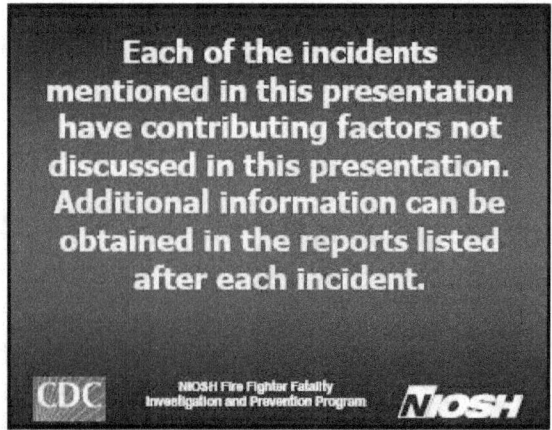
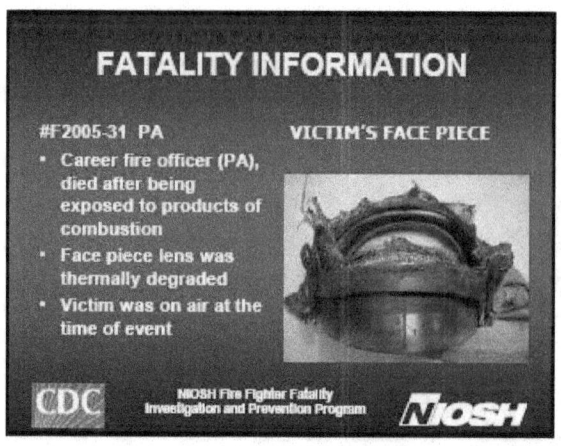
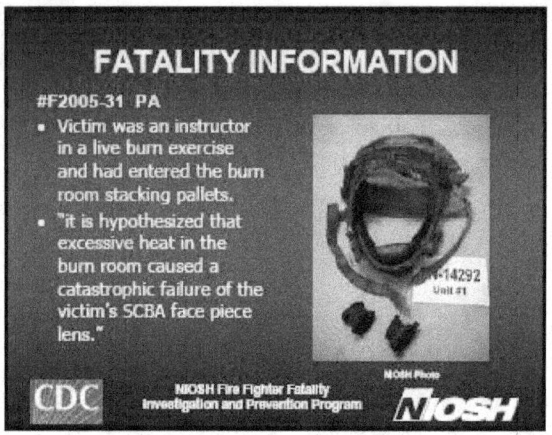
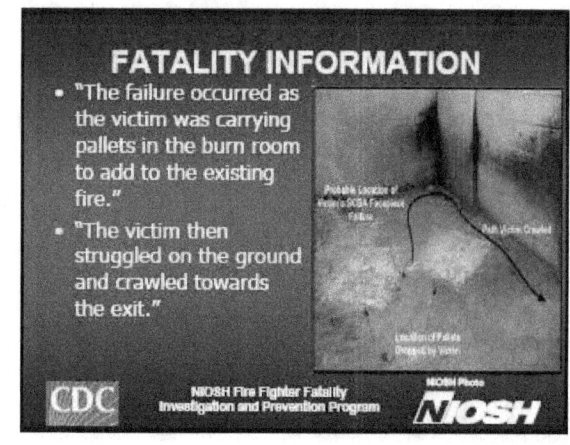
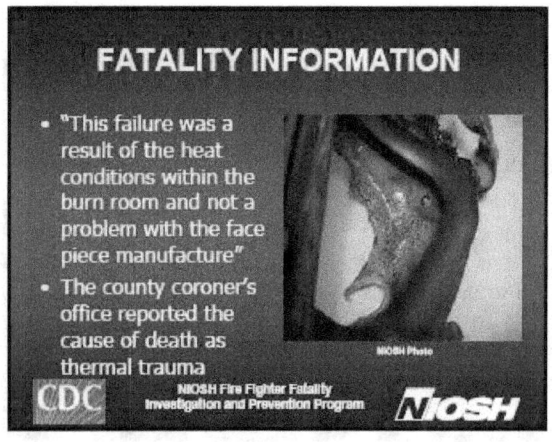

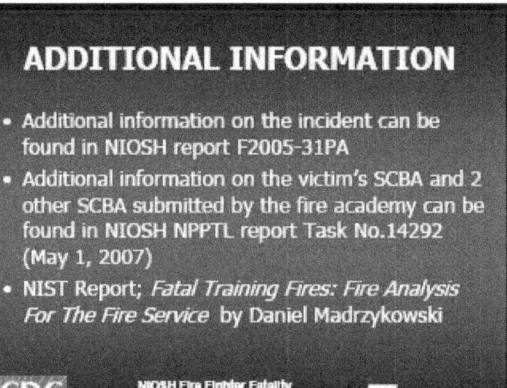

ADDITIONAL INFORMATION

- Additional information on the incident can be found in NIOSH report F2005-31PA
- Additional information on the victim's SCBA and 2 other SCBA submitted by the fire academy can be found in NIOSH NPPTL report Task No.14292 (May 1, 2007)
- NIST Report; *Fatal Training Fires: Fire Analysis For The Fire Service* by Daniel Madrzykowski

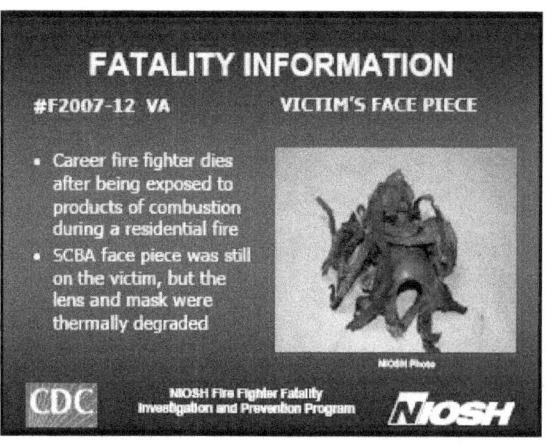

FATALITY INFORMATION

#F2007-12 VA — VICTIM'S FACE PIECE

- Career fire fighter dies after being exposed to products of combustion during a residential fire
- SCBA face piece was still on the victim, but the lens and mask were thermally degraded

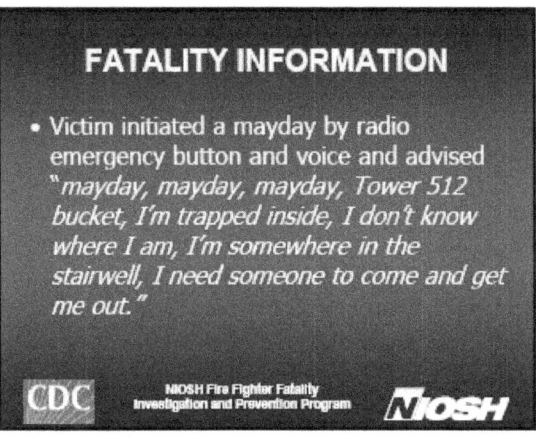

FATALITY INFORMATION

- Victim initiated a mayday by radio emergency button and voice and advised "*mayday, mayday, mayday, Tower 512 bucket, I'm trapped inside, I don't know where I am, I'm somewhere in the stairwell, I need someone to come and get me out.*"

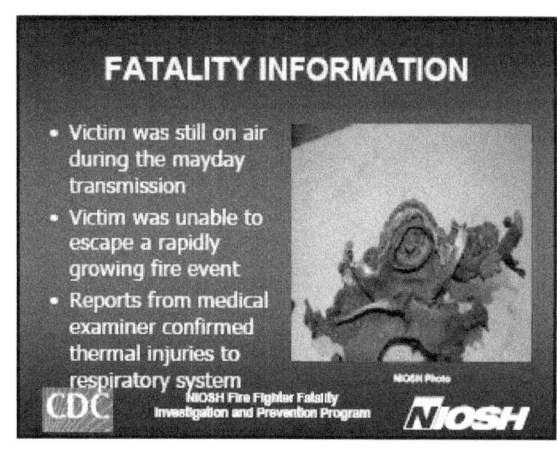

FATALITY INFORMATION

- Victim was still on air during the mayday transmission
- Victim was unable to escape a rapidly growing fire event
- Reports from medical examiner confirmed thermal injuries to respiratory system

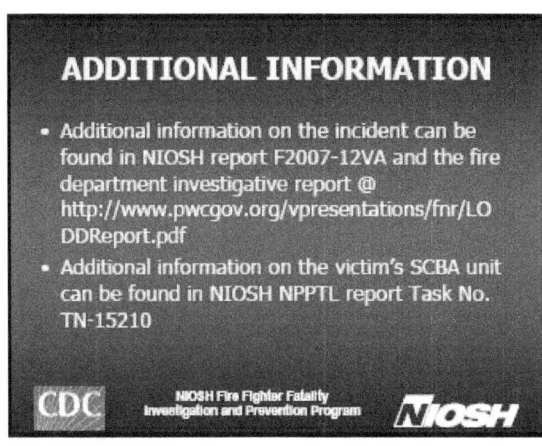

ADDITIONAL INFORMATION

- Additional information on the incident can be found in NIOSH report F2007-12VA and the fire department investigative report @ http://www.pwcgov.org/vpresentations/fnr/LODDReport.pdf
- Additional information on the victim's SCBA unit can be found in NIOSH NPPTL report Task No. TN-15210

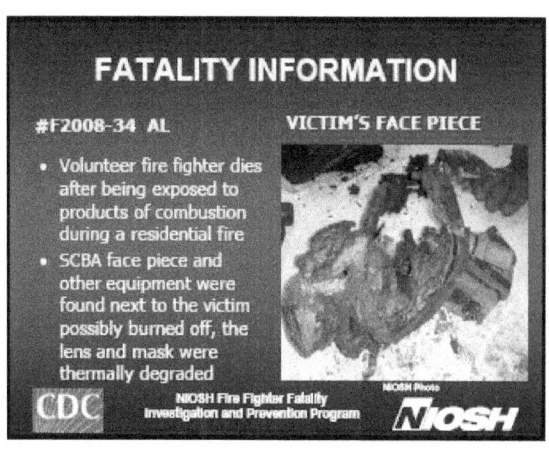

FATALITY INFORMATION

#F2008-34 AL — VICTIM'S FACE PIECE

- Volunteer fire fighter dies after being exposed to products of combustion during a residential fire
- SCBA face piece and other equipment were found next to the victim possibly burned off, the lens and mask were thermally degraded

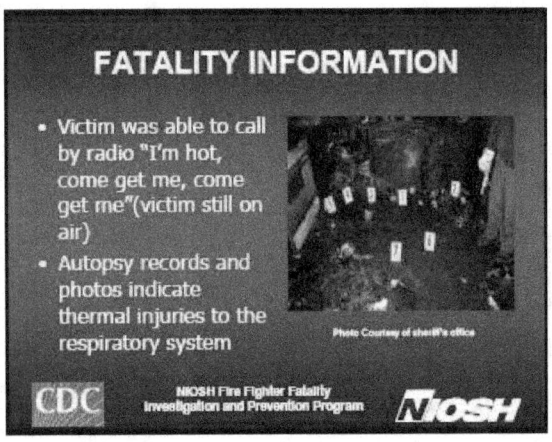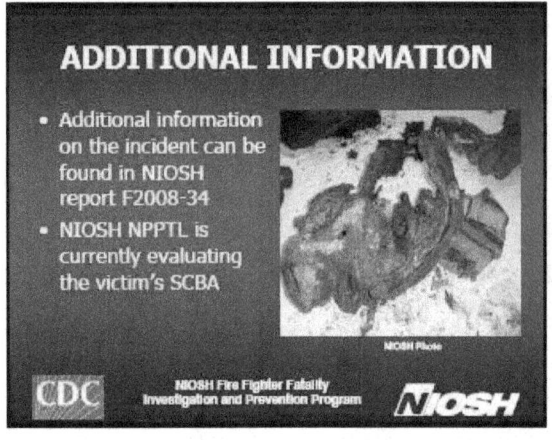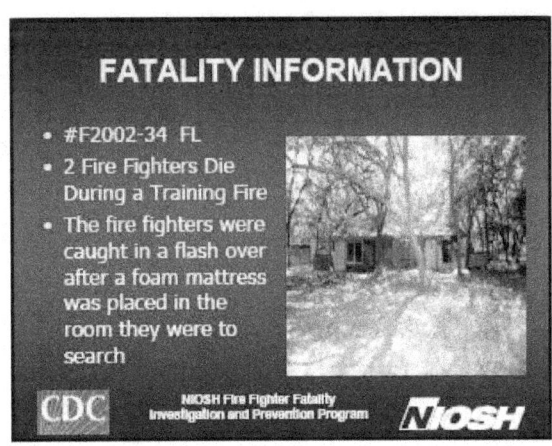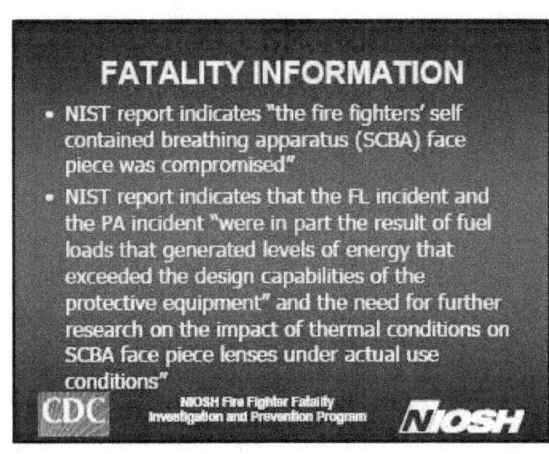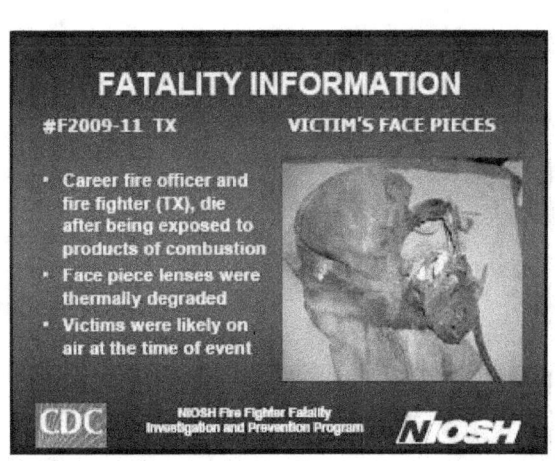

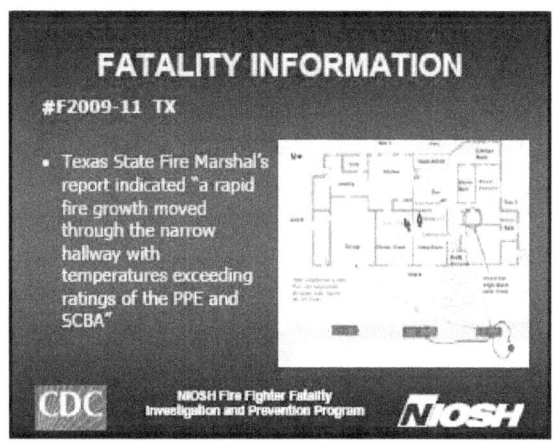
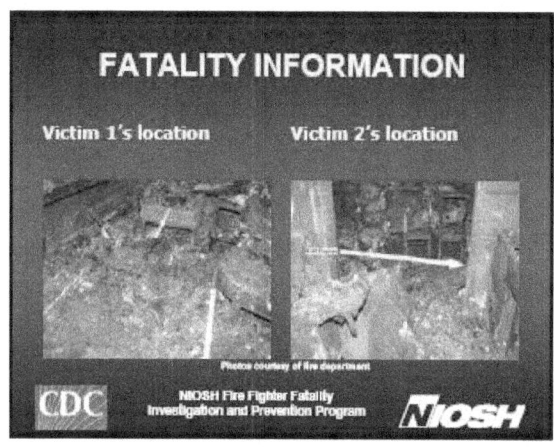
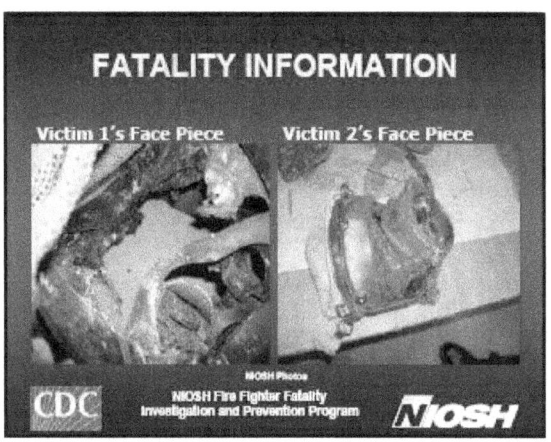
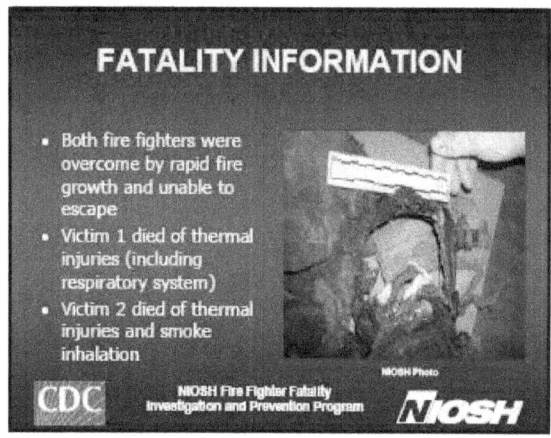
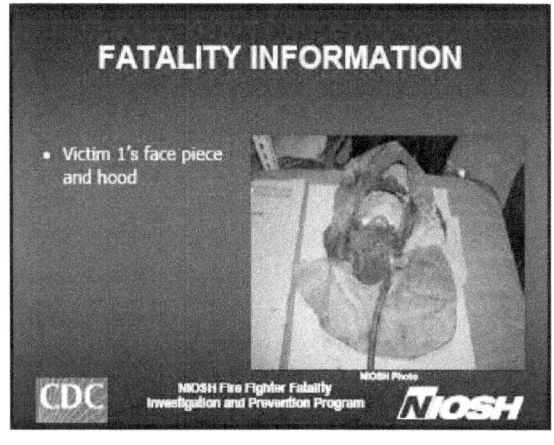
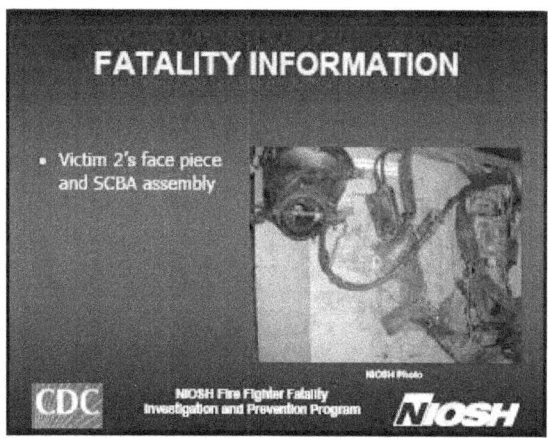

ADDITIONAL INFORMATION

- Additional information on the incident can be found in NIOSH report F2009-11TX
- Texas State Fire Marshal's Office, Investigative report number FY 09-01, Texas Department of Insurance, Austin Texas, April 12, 2009
- NIOSH NPPTL SCBA Status Investigation Report is still being developed
- NIST Fire Modeling is still being developed

Review of Findings From NIOSH FFFIPP Related to SCBA

- The fire fighter fatalities involved in the reviewed incidents were likely still "on air" at the time of the event
- Conditions encountered likely exceeded the performance capabilities of the SCBA face pieces
- Review of data suggest that the fire fighters suffered thermal injuries to the respiratory system
- Many of the fire fighters involved in the incidents were close to an escape point
- There were many other contributing factors in these FF fatalities that were not discussed in presentation

CONCLUSIONS

- Findings suggest a need for further research into the thermal resistivity capability of SCBA face piece
- Findings suggest a need to bring the present data forward to provide research bodies and other organizations with information to raise awareness and the protection levels of SCBA and PPE

NEXT STEPS

- NIOSH will continue to forward investigation reports from any future fire fighter fatality investigations that suggest performance issues with SCBA face piece or components that impact on NFPA certification
- We welcome suggestions from the work group on specific information that NIOSH could collect during fatality investigations

Thank You

www.cdc.gov/niosh/fire

This information is distributed solely for the purpose of pre-dissemination peer review under applicable information quality guidelines. It has not been formally disseminated by the National Institute for Occupational Safety and Health. It does not represent any agency determination or policy.

APPENDIX 3.F – Massachusetts Fire Training Academy
Fred LeBlanc, Massachusetts Fire Training Academy

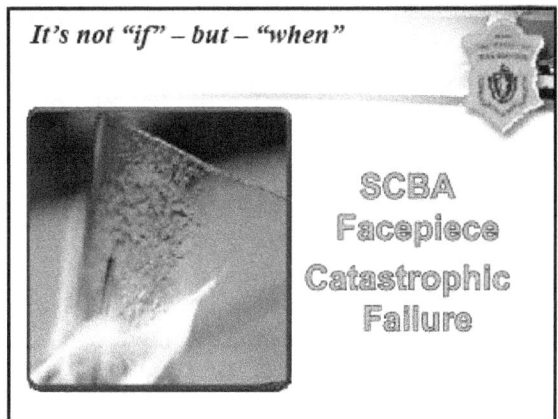

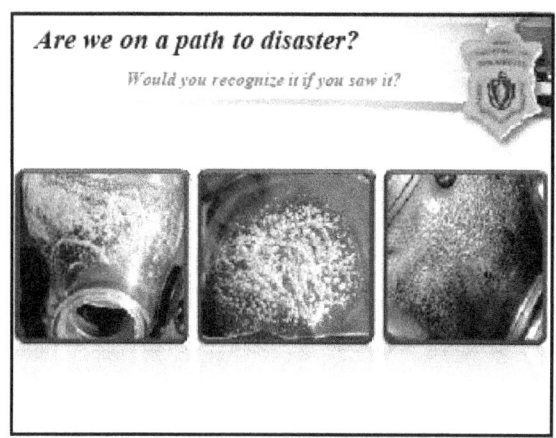

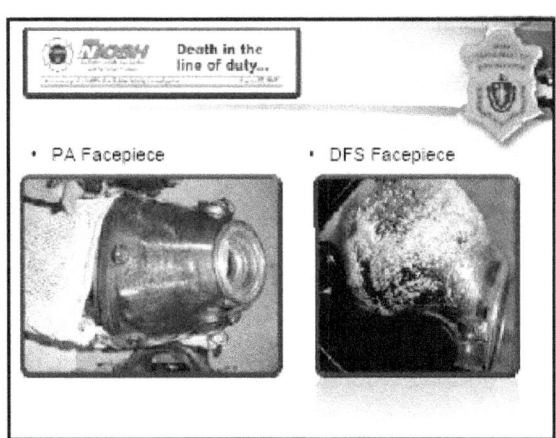

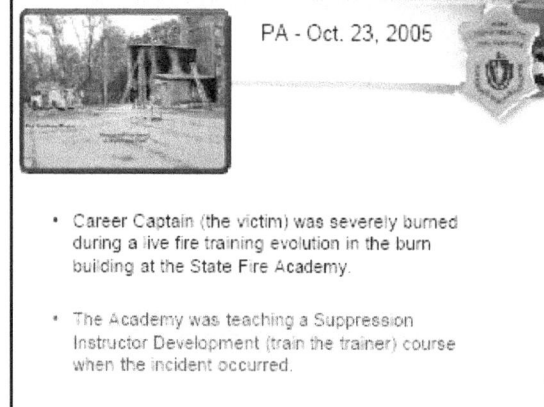

PA - Oct. 23, 2005

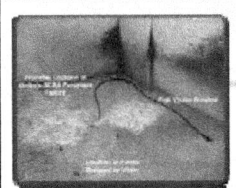

- Victim was loading pallets in basement then came to top of stairs pulling and lifting his gear.
- Second instructor asked "are you OK". Victim replied "It's hot as hell down here."
- Second instructor stated "you really need to go outside."
- Victim replied "No, I'm all right. Got up and said,"Yeah I'm fine I'll see you down there."

PA - Oct. 23, 2005

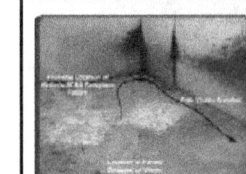

- The victim was transported via ambulance to a community hospital were he was stabilized prior to transport via helicopter to a regional trauma/burn center. The victim died from his injuries on October 25, 2005.
- Heat within burn room caused catastrophic failure of mask
- Not a problem with manufacture

NIOSH Recommendations

- Ensure that two training officers are present with a charged hoseline during the ignition or refueling of a training fire in accordance with NFPA 1403.
- Determine the minimum amount of flame, heat and/or smoke required during live fire evolutions to perform the training while ensuring fire fighter safety.
- Use the minimum fuel load necessary to conduct live fire training.

NIOSH Recommendations

- Have a written respiratory protection program and ensure that self-contained breathing apparatus (SCBAs) facepieces are properly inspected, used, and maintained.
- Have burn rooms with at least two exits.
- Avoid having basement burn rooms

Additional Recommendations

- Installing instrumentation within live fire training structures to record information such as heat, the effects of suppression and the byproducts of combustion.
- Installing a ventilation system within the burn structure.
- Having a qualified engineer evaluate fuel loads, heat retention, and the instrumentation and ventilation systems of live fire training facilities.

Temperature – Time - Result

Lab Test

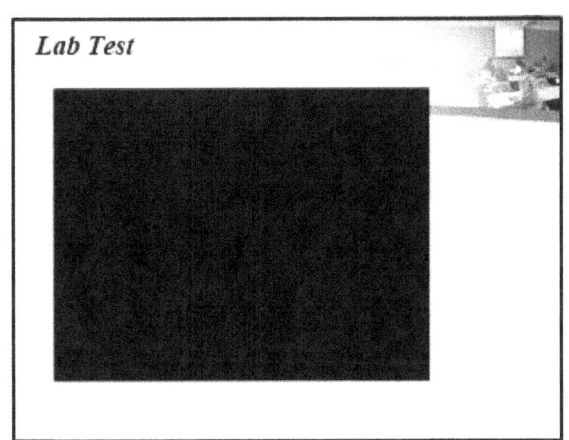

Consequence

- 690 Degrees F.

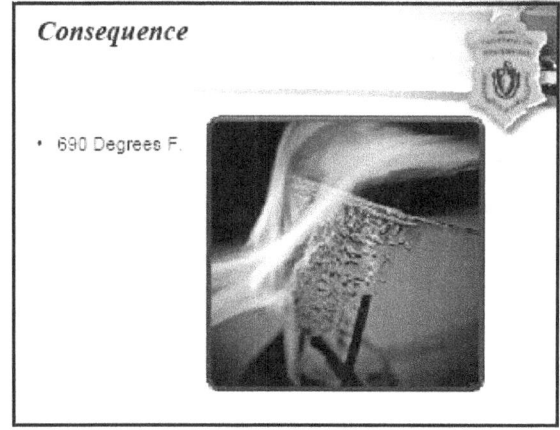

Lab Test

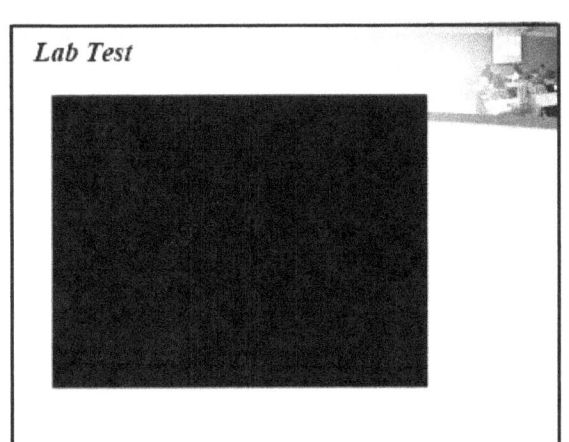

Were here - Why we need to change

- We have always done it this way
- This way is more efficient
- We deal with limited time and help
- You don't understand

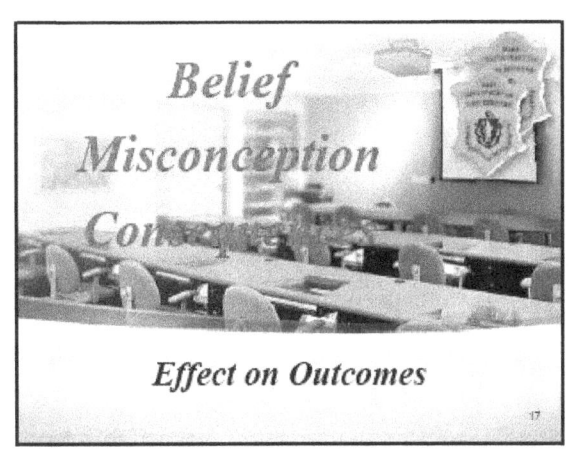

Belief Misconception Consequence

Effect on Outcomes

Difference in Decision Making

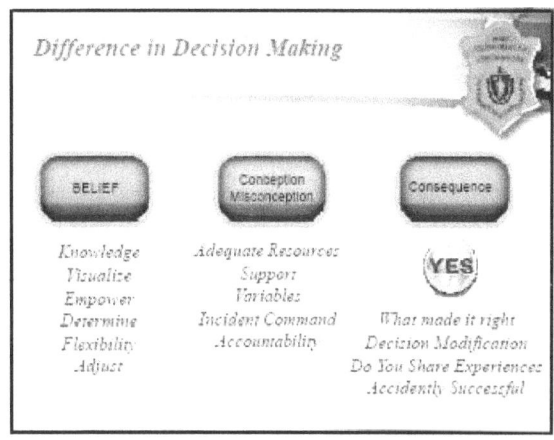

BELIEF	Conception Misconception	Consequence
Knowledge	Adequate Resources	YES
Visualize	Support	
Empower	Variables	What made it right
Determine	Incident Command	Decision Modification
Flexibility	Accountability	Do You Share Experiences
Adjust		Accidently Successful

Difference in Decision Making

BELIEF	Conception Misconception	Consequence
Knowledge	Adequate Resources	
Visualize	Support	NO
Empower	Variables	
Determine	Incident Command	What made it wrong
Flexibility	Accountability	Decision Modification
Adjust		Do You Share Experiences

Belief & Misconception

- Belief – Pulling a dry line into a burning structure is efficient.
- Belief – When the low air alert sounds I have plenty of air to exit safely.
- Belief – My health is my business.
- Belief – Slowing down at intersections is good enough.

Phase Fires

- Objectives of the phase
- Teaching points
- Which apparatus are being used
- Building layout
- Squad rotation schematic

Bale Requirements

- What are the requirements for each phase
- What are requirements for other programs such as SFFP, Flashover, & Call/Volunteer Programs
- Limitations in Springfield or other authorized locations
- Behavior of burn building interior tiles

Constructing Fires

- Coordinate ignitors escape route
- Meet at predetermined location
- Designated radio channels
- Drivers routes
- ¼ bale for smudge fires
- Small fire located away from main fire
- Starter fire lit after smudge fire
- Starter fire slid under rack
- Timing must be optimum

Instructional Staff

- Protective clothing purchases
- SCBA maintenance
- Effects of sweating & weather
- Change gear
- Need for scanning radios
- Coordination with IC, Inside Safety, Outside Safety
- THANKS for doing things RIGHT!

APPENDIX 3.G – NIST Respirator Research

Nelson Bryner, National Institute of Standards and Technology

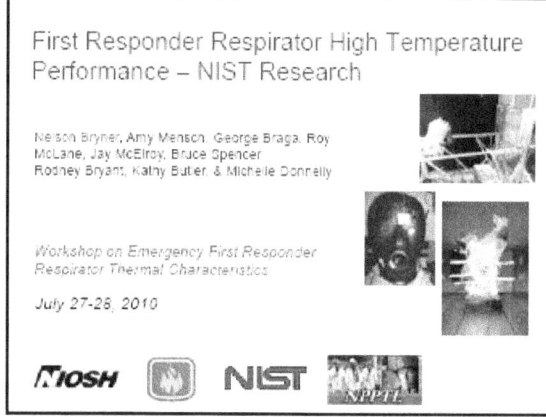

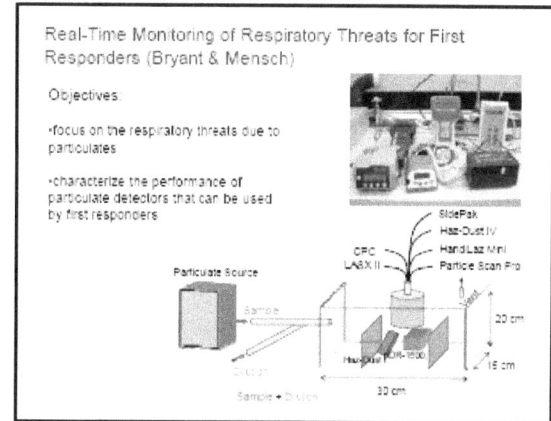

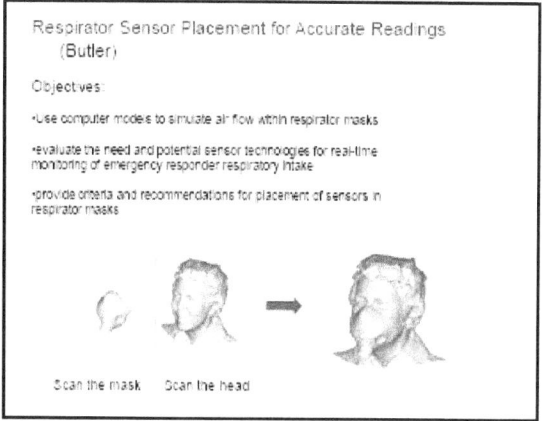

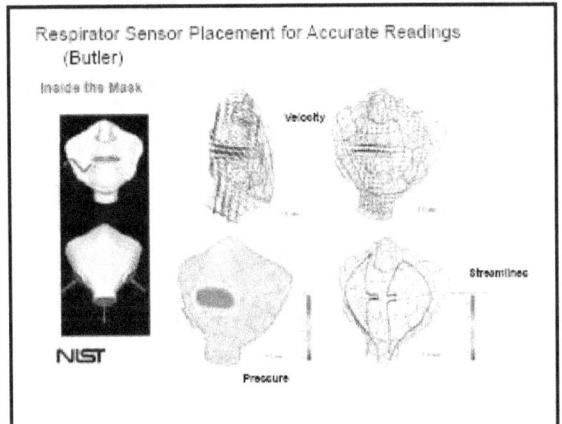
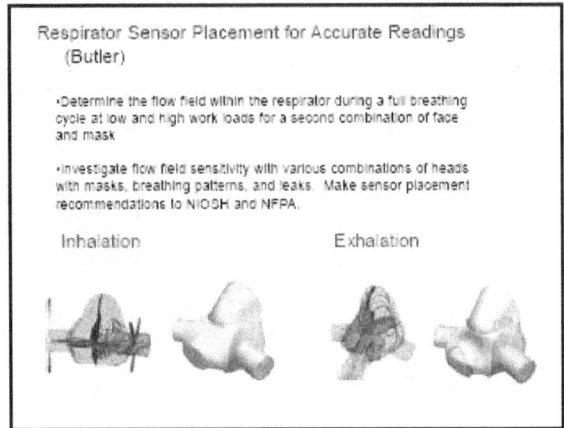
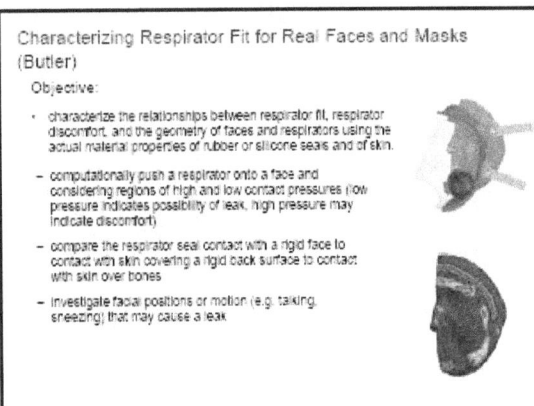
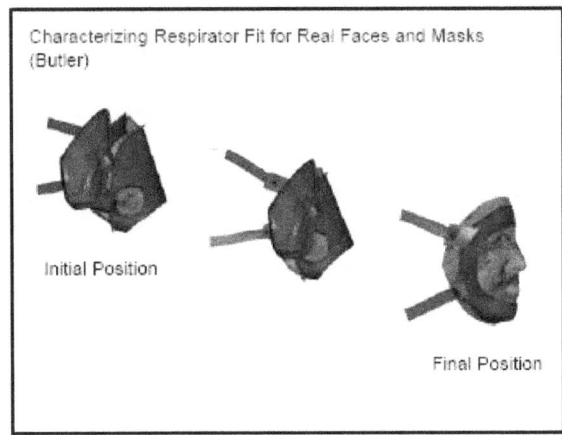
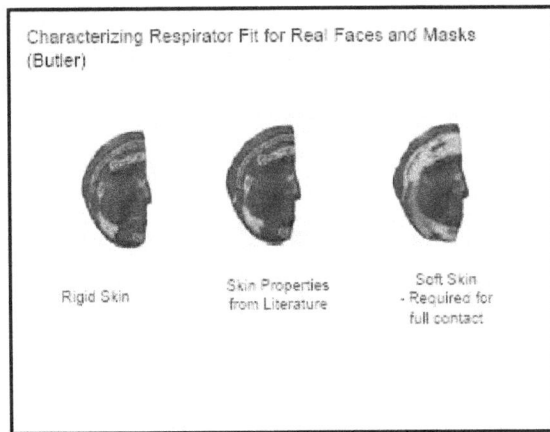
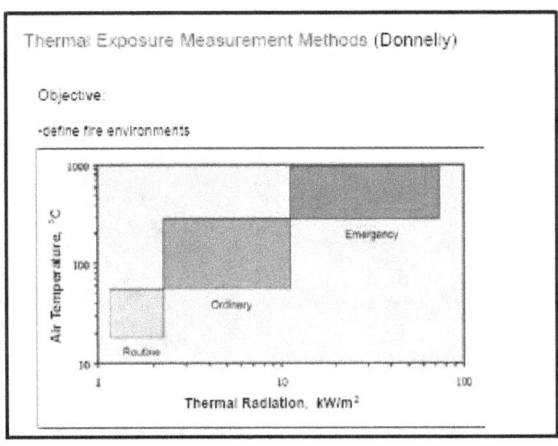

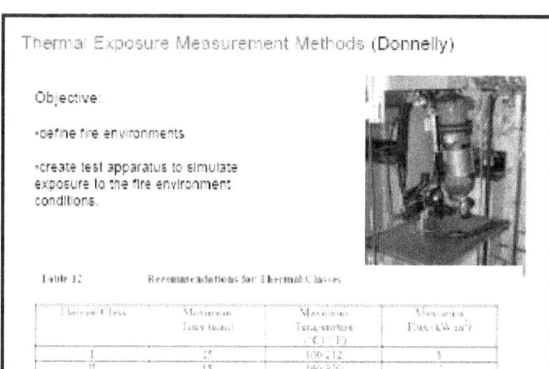
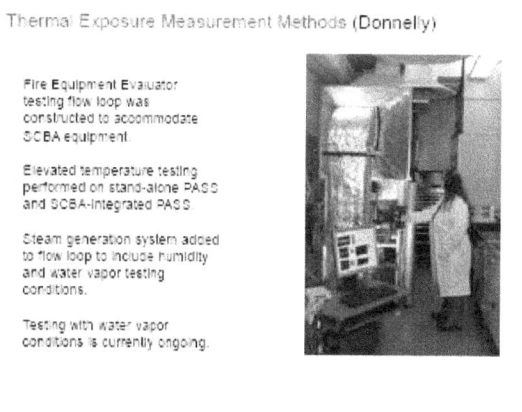
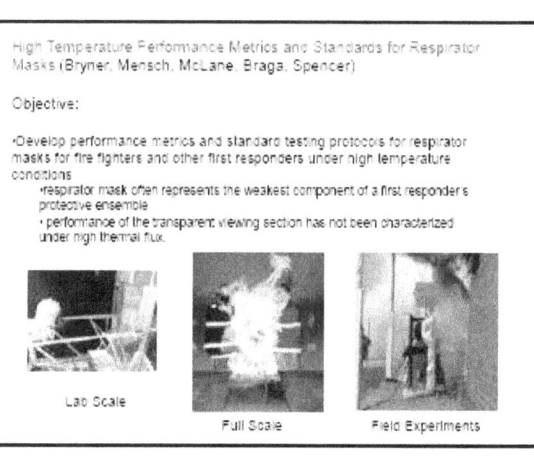
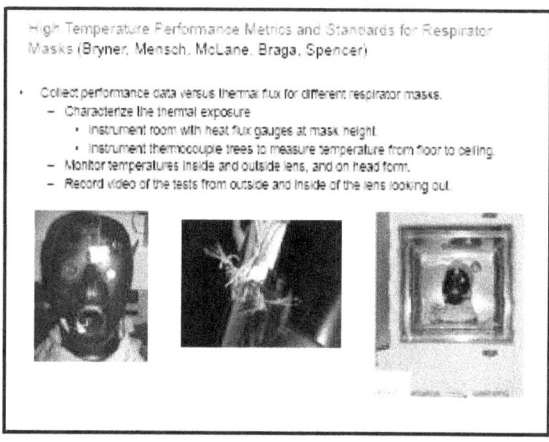
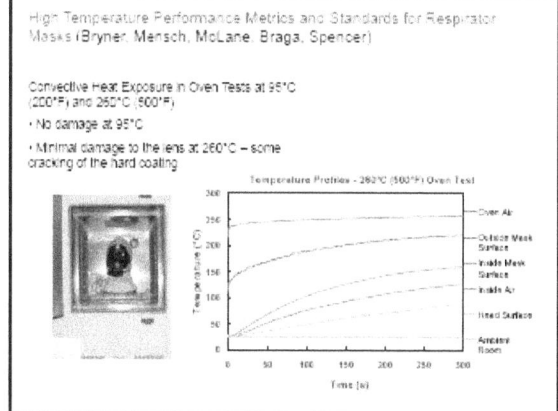
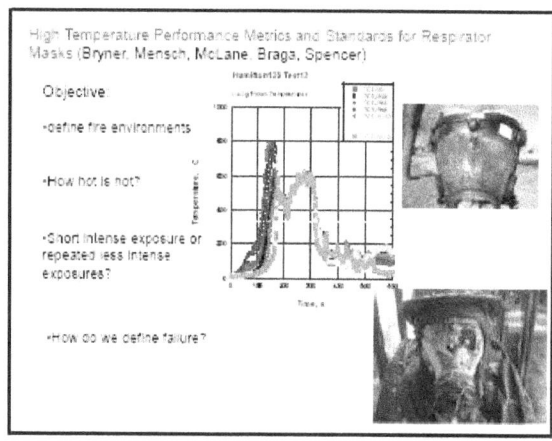

High Temperature Performance Metrics and Standards for Respirator Masks (Bryner, Mensch, McLane, Braga, Spencer)

Full Scale Field Experiments
- Fire tests in 2-story townhouses, suppressed approximately 5 minutes after ignition.
- Placed masks in the kitchen, 0.8m (31in) above the floor.

High Temperature Performance Metrics and Standards for Respirator Masks (Bryner, Mensch, McLane, Braga, Spencer)

Full Scale Room with Burner Experiments
- 12'x12' room with natural gas burner on the floor
- Exposed almost 100 masks to different conditions
  - Breathing w/ humidity / Non-breathing
  - 2 kW/m² to 20 kW/m² heat flux
  - Radiative / Convective

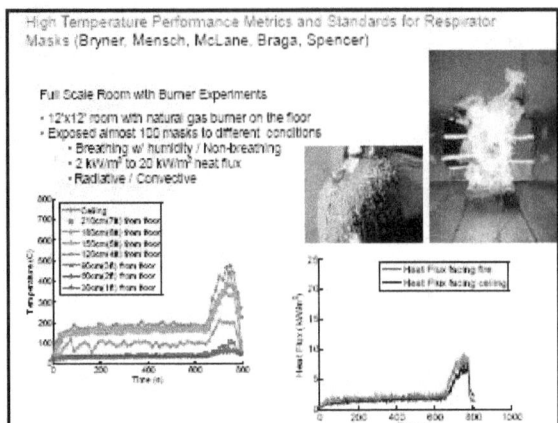

High Temperature Performance Metrics and Standards for Respirator Masks (Bryner, Mensch, McLane, Braga, Spencer)

Future Work

- Extend performance data vs. thermal flux to update standards organizations on radiative vs. convective standard test (NFPA & ASTM)
- Examine the performance of the masks during cyclical heated exposures vs. a single extreme exposure
- Characterize the chemical/physical mechanisms causing the cracking, crazing, bubbling (failure modes)
- Develop new lens/respirator mask technology to enhance thermal protection of respiratory protection

Where did 200 F, 15 minutes, and 10 seconds originate?

An alternative thermal classification of structural fires was developed by the International Association of Fire Fighters (IAFF) under Project FIRES. As above, fires can be classified by their temperature and rate of heat output. In addition, each class of fire can be associated with a structural fire fighting situation and its expected average duration as listed below.

- *Class I* occurs in a room during overhaul. Environmental temperatures up to 200°F (311°C) and thermal radiation up to 0.05 watts/cm² (0.012 cal/cm²s) are encountered for up to 30 minutes.

- *Class II* occurs when a small fire is burning in a room. In this case environmental temperatures from 100°F to 200°F (311°C to 367°C) and thermal radiation from 0.050 to 0.100 watts/cm² (0.012 to 0.024 cal/cm²s) are encountered up to 15 minutes.

- *Class III* exists in a room that is totally involved. Environmental temperatures from 200°F to 500°F (367°C to 533°C) and thermal radiation from 0.100 to 0.175 watts/cm² (0.024 to 0.042 cal/cm²s) are encountered up to 5 minutes.

- *Class IV* occurs during a flashover or backdraft. Environmental temperatures from 500°F to 1500°F (533°C to 816°C) and thermal radiation from 0.175 to 4.2 watts/cm² (0.042 to 1.0 cal/cm²s) are encountered up for *approximately* 10 seconds.

Abeles, Fred J., "Protective Ensemble Performance Standards, Project Fires" Phase 1B Final Report, Volume 2, Grumman Aerospace Corporation, Bethpage New York, May 1982.
Duffy, R. M., Danker, J. C., Baer, A. R., "Project Fires, Firefighters Integrated Response Equipment System – The Final Report," International Association of Firefighters, Department of Occupational Health and Safety, Washington D.C., 1982.

APPENDIX 4 – WORKING GROUP QUESTIONS

NIOSH, NIST, NFPA/FPRF
Workshop on Emergency First Responder Respirator Thermal Characteristics

Location: NIOSH NPPTL, **Building 140, 626 Cochrans Mill Road, Pittsburgh, PA**
Date: **27-28 July 2010**

WORKING GROUP QUESTIONS
Last Updated: 12 July 2010

Each of the working groups should individually address the following set of specific topics/questions, and report back to the whole group:

1) Identification of Performance Needs (Day 1)
    a. Current Equipment
        1. What components of the overall SCBA package require attention, and in what priority order (e.g. face piece, straps, connectors, PASS, etc)?
        2. Are there certain repeatable sequences of operation that magnify or promote equipment failure (e.g. repeated training exposure with minimal cool-down, long shelf life, winter usage, etc)?
        3. What parameters should be used to define the realistic limits of operability (e.g. temperature exposure, visibility thru face piece, etc)
    b. Current Practice and Usage
        1. What are the prioritized conditions of use that are of most concern (e.g. certain structural fires, training, seasonal conditions, etc)?
        2. What specific operational concerns, if any, need to be addressed (repeated high temperature exposure, pre-failure indicators, replacement protocols, etc)?
        3. What specific training concerns, if any, need to be addressed (recognition of failure markers, maintenance practices, etc)?
    c. Future Trends
        1. How are SCBA expected to change based on current technological trends (e.g. equipment, operation, training, etc)?
        2. What perceived problems might be anticipated with future SCBA (e.g. adaptation of support infrastructure, physiological complications, etc)?
    d. Other Issues
        1. What are known or potential topics of technical debate (e.g. method to evaluate face piece integrity, etc)?
        2. What specifically needs to be addressed from a regulatory or standardization standpoint?
        3. What single message should the fire service express on this topic in terms of performance needs?

2) **Development of Research Priorities (Day 2)**
    a. **Short Term Research Needs**
        1. What are short term research priorities for equipment (e.g. improved face piece materials, defining PASS signals operation, etc)
        2. What are short term research priorities for usage and practice (e.g. survey of field usage, establishing evaluation protocols, etc)
    b. **Long Term Research Needs**
        1. What are long term research priorities for equipment (e.g. development of evaluation test methods, identifying real-time measurement techniques, etc)
        2. What are long term research priorities for usage and practice (e.g. defining culture of use, establishment of realistic physiological benchmarks, etc)
    c. **Other Issues**
        1. What constituent groups and/or organizations need to be involved (e.g. clinical/physiological, materials science, etc)?
        2. What single message should the fire service express on this topic in terms of research priorities?

APPENDIX 5 – WORKING GROUP RAW RESULTS

APPENDIX 5.A – Working Group A Raw Results

Table 3 - Group A Identification of Performance Needs Summary

Current Equipment	Current Practice and Usage
<ul><li>Face piece is weakest component</li><li>Straps fraying</li><li>Parameters to define limits of operability of lens<ul><li>Optically clear</li><li>Impact</li><li>Abrasion resistant</li><li>Distortion requirement</li><li>Thermal performance</li></ul></li><li>PASS performance – multiple tones/failure to recognize</li><li>Thermal warning device</li><li>Temperature display</li><li>Ensemble Testing</li></ul>	*Conditions of Use*<ul><li>Imminent hazard indicator – temperature, heat flux or rate of rise of temperature indicator</li><li>Duration and intensity of exposure<ul><li>How long can a fire fighter survive?</li></ul></li><li>Representative fire exposure<ul><li>300 °F to 400 °F minimum in hallway</li><li>500 °F</li><li>>1200 °F to 1500 °F during flashover/in a room with fire</li></ul></li><li>Survivable fire exposure</li><li>80kW/m^2 TPP test</li><li>Radiant/Convective</li></ul>
Future Trends	*Operational Concerns*<ul><li>Temperature range extremes: hot to cold, cold to hot</li><li>Nose cup deflecting cool air to exhaust</li><li>Soot layer on the lens</li><li>Simultaneous exposure to heat, impact and rub tests</li></ul>*Training Concerns*<ul><li>Insufficient quantity of live fire training</li><li>Inexperienced fire fighters -losing a generation</li><li>Loss of generational experience</li><li>Fewer fires today</li><li>Promoted from EMS to fire captains</li></ul>
<ul><li>Glass lens</li><li>Composite lens</li><li>Flat lens geometry (watch glass – quartz)<ul><li>Adequate thermal props</li></ul></li><li>Two part Acme goggles</li><li>NASA space shuttle windows</li><li>Peel away sacrificial layer</li><li>Provide additional cues to hazard<ul><li>PPE prevents sense of hazard</li></ul></li><li>Bimetallic strip alarm</li><li>Temp indicator</li><li>Radiant heat/temperature indicator<ul><li>Worse case algorithm</li><li>Audible alarm (FF + IC)</li></ul></li><li>Reflective Coating</li><li>Loss of Experience</li><li>Flat Pack<ul><li>Wear into confined space</li><li>Decreased physiological burden</li><li>Less flexible when charged</li><li>Less effort - FF can spend more time in fire</li></ul></li><li>Different ratings for masks<ul><li>Confined space</li><li>Fire</li><li>Non-fire</li></ul></li></ul>	**Other Issues**<ul><li>Uniform testing of entire ensemble</li><li>Consistent with environment FF will encounter 500 °F</li><li>More Realistic radiant exposure test</li><li>Trade off visibility/better performance</li><li>Standard and SCaM documents</li></ul>

Table 4 - Group A Development of Research Priorities Summary

Short Term Research Needs	**Long Term Research Needs**

• Lens – improve material
• Communications
• Interface with SCBA
• Representative scenarios where lens are not performing
   • trapped / lost FF
   • daily or routine use
• ID thermal stresses - use to improve testing methods
• Adhere to best practices
   • Behavior
   • Training program
• Quantify usage areas
   • Peer review – best practices
   • kW/m^2, temp, time
• Temp Indicators / displays
   • TIC or HUD
• Unified performance requirements between SCBA and PPE
• Define fire environment
   • Radiant / convective - develop appropriate test protocols
	• Lens system - improve materials - implementation 15 years
	• Communications - crew / IC
	• Technology integration
	   • Information overload
	• Enhanced standard uniformity
	   • Uniform performance levels

Other Issues	**Single Message Expressed**

• Funding for lens replacement
   • Fire Grants vs. SCBA main. budgets
• Department involvement -
   • Knowledge gap
   • Information dissemination
• Engage service organizations IAFC, IAFF NVFC, NIOSH etc.
   • Fire service media
• Regulation-
   • NIOSH advisory to replace?
• Address adherence to best practices
   • Behavior
   • National and local
	• Unified equipment performance criteria
	   • representative & realistic
	• Failure point should not result in death
	   • weak link should not be respiratory
	• Funding Resources
	   • Information dissemination
	   • Training
	   • Equipment

APPENDIX 5.B – Working Group B Raw Results

Table 5 - Group B Identification of Performance Needs Summary

Current Equipment	**Current Practice and Usage**

• Need to have approach that considers entire ensemble (For both severity and frequency)
• Need inherent built-in methods for "warning"
• Need to define environment
  • Normal vs. ordinary vs. emergency (where we want "survival")
  • What levels of protection do we want
• Need to clarify product life cycle
• Need to identify factors that cannot be compromised (e.g. transparency, impact resistance) in addition to thermal characteristics
• Need to consider not only equipment enhancement, but also technology for other purposes (e.g. early and reliable warning system)
• Feedback from the field is that the mask is the current weak link
• Consider alternative materials (e.g. layering of materials)
• Not practical to lower current protection of other gear. Improving mask protection is the practical option
  • Exception: Mandating an upper limit for TPP (possibly for certain components like hoods or gloves)
• Consider impact of added face piece accessories used with current materials
• Design to prevent "catastrophic failure"
• Repeatable sequence is "rapid fire growth"
• Data is lacking and needed on repeatable exposure and the long term effect
• Establish priority of performance characteristics
  • Impact (catastrophic consequences)
  • Thermal Resistance (catastrophic)
  • Abrasion Resistance
  • Transparency
  • Cost
  • Life cycle/durability
  • Some are more convenience by still can lead to injuries and fatalities indirectly.
	• This is a behavioral and training issue
	• Need straight-forward signal that users can easily relate to (e.g. measuring rate of temp. change)
	• Current fire service environment is highly variable
	• Design for protection needed in room next to room that flashed
	• Need better understanding (training) of gear care and maintenance

	Future Trends
	• Consider special equipment specifically designated for flashover training • Consider additional flip-down lens or other supplemental lens • Consider active cooling system • Provide built-in warning methods • Measurement of inhaled air temperature • Might have to look at minimizing weight
	Single Message Expressed
	• All ensemble elements should be equivalent in performance • Extend survivability time in emergency conditions. • Optimize best use of existing materials through reevaluation using existing and new test methods • Must address behavior, usage, training, and education.

Table 6 - Group B Development of Research Priorities Summary

Short Term Research Needs	Long Term Research Needs

• Identify Training Needs
• Identify flux level for lens failure
   • Mechanical (can have engineered solution)
   • Human (e.g. Training)
   • Combination of both
• Identify mechanical solutions for lens failure
• Identify human-based solutions for lens failure
• Evaluate existing test methods (e.g. flux, temp, time, etc.)
• Collection of credible data on actual field usage (not only LODD events)
• Clarify impact of repeated exposure (i.e. Training vs. field ops)
• Establish pro-active approach to use current info and best practices for training programs (thru NAFTD-North American Fire Training, USFA-NFA, etc.)
	• Address training needs
	• Hold "Project Fires" part II with a comprehensive overview of issue to establish target usage (w/ meta-analysis) (e.g. national/international involvement)
	• Revise or develop new test methods
	• Establish framework and protocol for long term data collection needs (e.g. PASS device, autopsy, etc)
	• Holistic ensemble testing (e.g. develop ensemble TPP thresholds, min/max)
	• Development of active warning system
	• Development of new materials, designs, and approaches (e.g. active cooling)
	• Establish research clearinghouse

Other Issues	**Single Message Expressed**

• Groups/organizations needed (and not currently involved
   • Academia, DOD, NASA, USFS, etc.
   • International
   • Other technical committees (e.g. NFPA 1500, 1403, 1971, 1981, 472, and 1800, ASTM, ISO)
	• Determining flux, temperature and time is critical
	• Address training needs
	• Evaluate current state of the art, educate approach for improvement, enhance design, materials, equipment to improve survivability

APPENDIX 5.C – Working Group C Raw Results

Table 7 - Group C Identification of Performance Needs Summary

Current Equipment	**Current Practice and Usage**

• Lens is number 1 priority, followed by the face piece, and the rest of the SCBA
• Complete ensemble testing
• Pressure boundaries should be maintained
• Training and behavioral issues contribute to equipment failure
• Surviving a catastrophic event is critical
   • Should survive 1 flashover
• Mask temperature does not equal environment temperature
	• All situations are important to design for
	• There is no such thing as routine fire fighting
	• High temperature exposure is operational concern
	• SCBA should be tested as an ensemble
	• Use the HUD to monitor the temperature of the lens for pre-failure indicator
	• The same mask should be used for training and fire fighting
	• Training can help change the fire service culture, which values dangerous "macho" behavior
	• Enforcement of training is difficult
	• Several points to emphasize in training involving personal responsibility
	   • Visual inspection of equipment
	   • Don't walk into the fire
	   • Follow best practices
	• Multiple agencies partner on a video, similar to Houston's demonstration, showing the limits of the equipment to distribute and use in training
	   • Videos would have high impact

Future Trends	**Other Issues**

• Should we go backwards and reduce/limit the level of protection in the rest of the equipment?
   • Fire fighter could better sense the danger of the environment and acted cautiously
• Use new materials
   • Polyethersulfone has a heat deflection temperature about 100 °F higher than polycarbonate, and is almost as transparent
   • Gold coating to reflect heat
• Changes to the standards are the effective way to push technology to advance
• Test parameters higher/lower
• Will another component just become the weakest link?
• Would a new SCBA still fail in a catastrophic event?
	• Face piece/lens test
	• A wipe/impact test that occurs after conditioning
	• Complete ensemble testing
	• ANSI-Z87.1 transparency test may not be passable by new materials
	• Single message would be to address the training/culture of the fire service in an "in your face" manner with a video created by a joint effort with NIOSH, NFPA, IAFF, IAFC, IFSTA, DELMAR, NVFC, NIST

Table 8 - Group C Development of Research Priorities Summary

Short Term Research Needs	**Long Term Research Needs**

• Research new material (polyethersulfone)
   • Other 3rd party evaluation (NIOSH, NIST, other)
   • Test visual acuity
   • Test thermal properties
   • Test impact/abrasion resistance
• Define heat flux/temp levels for survivability
• Replicate "real world" conditions in reproducible lab tests
   • What are the fire conditions?
   • What are conditions immediately after event?
• Produce and evaluate a survey/questionnaire after incidents where masks are damaged, for gathering information
   • Need to determine what information needs to be gathered
   • Should be short, 4-5 questions, and have pictures to show examples of damage
	• Create a temperature tag on the respirator to gather real time data
	• New test method
	   • heat exposure
	   • heat/flame
	• Long term: use/practice
	   • Training tools tailored to the "new generation" of fire fighters
	   • You-tube, video games, etc.

Other Issues	**Single Message Expressed**
• Groups to be involved • Technology/Training warehouse • Responder.gov clearinghouse • Survivability training for fire fighters, simulators, similar to what the military uses	• There are limitations of the PPE, and this must be stressed in training • Discipline is a related issue

www.ingramcontent.com/pod-product-compliance
Lightning Source LLC
Chambersburg PA
CBHW081738170526
45167CB00009B/3867